Praise for David Lindley's

UN**CERTAINTY**

"Layers keen human drama on top of mind-bending scientific advancement." —*Discover Magazine*

"A well-written account. . . . [With] good old-fashioned narrative and quirky characters."
—*The Providence Journal*

"This story has been told before but seldom with such clarity and elegance." —*Scientific American*

"Impressively researched." —*New Scientist*

"A physicist and skilled science writer, Lindley neatly sketches the players and chessboard at the Solvay Conferences, where Einstein lost his battle against the quantum world." —*USA Today*

"Lindley captures the passion of the struggle, showing both the public controversies and the sometimes harsh private judgments. . . . The story is told with verve." —*Nature*

DAVID LINDLEY

UNCERTAINTY

David Lindley holds a Ph.D. in astrophysics and has been an editor at *Nature, Science,* and *Science News.* He is the author of *The End of Physics, Degrees Kelvin, Where Does the Weirdness Go?,* and *Boltzmann's Atom.* He lives in Alexandria, Virginia.

UNCERTAINTY

UN**CERTAINTY**

*Einstein, Heisenberg, Bohr,
and the Struggle for the
Soul of Science*

DAVID LINDLEY

Anchor Books
A Division of Random House, Inc.
New York

FIRST ANCHOR BOOKS EDITION, FEBRUARY 2008

The Library of Congress has cataloged the Doubleday edition as follows:
Lindley, David, 1956–
Uncertainty : Einstein, Heisenberg, Bohr, and the struggle for the soul of science / David Lindley. — 1st ed.
p. cm.
Includes bibliographical references.
1. Heisenberg uncertainty principle. 2. Physics—Philosophy. I. Title.
QC174.17.H4L56 2007
530.12—dc22
2006017029

Anchor ISBN: 978-1-4000-7996-4

Author photograph © Hellen Gelband
Book design by Michael Collica

He is the God of order and not of confusion.
— *Isaac Newton*

Chaos was the law of nature; order was the dream of man.
— *Henry Adams*

CONTENTS

UNCERTAINTY

INTRODUCTION

If science is the attempt to extract order from confusion, then in early 1927 it veered onto an unexpected path. In March of that year, Werner Heisenberg, a physicist only twenty-five years old but already of international renown, set down a piece of scientific reasoning that was in equal measure simple, subtle, and startling. Heisenberg himself could hardly claim he knew exactly what he had done. He struggled to find an apt word to capture the sense of it. Most of the time he used a German word readily translated as "inexactness." In a couple of places, with a slightly different intention, he tried "indeterminacy." But under the irresistible pressure of his mentor and sometime taskmaster Niels Bohr, Heisenberg grudgingly added a postscript that brought a new word onto the stage: *uncer-*

tainty. And so it was that Heisenberg's discovery became indelibly known as the uncertainty principle.

It's not the best word. Uncertainty was hardly new to science in 1927. Experimental results always have a little slack in them. Theoretical predictions are only as good as the assumptions behind them. In the endless back-and-forth between experiment and theory, it's uncertainty that tells the scientist how to proceed. Experiments probe ever finer details. Theories undergo adjustment and revision. When scientists have resolved one level of disagreement, they move down to the next. Uncertainty, discrepancy, and inconsistency are the stock-in-trade of any lively scientific discipline.

So Heisenberg didn't introduce uncertainty into science. What he changed, and profoundly so, was its very nature and meaning. It had always seemed a vanquishable foe. Starting with Copernicus and Galileo, with Kepler and Newton, modern science evolved through the application of logical reasoning to verifiable facts and data. Theories, couched in the rigorous language of mathematics, were meant to be analytical and precise. They offered a system, a structure, a thorough accounting that would replace mystery and happenstance with reason and cause. In the scientific universe, nothing happens except that something makes it happen. There is no spontaneity, no whimsy. The phenomena of nature might be inordinately complicated, but at bottom science must reveal order and predictability. Facts are facts, laws are laws. There can be no exceptions. The mills of science, like those they replaced, would grind exceeding small. And just as perfectly.

For a century or two, the dream seemed realizable. If scientists of one generation, building on the work of the last, could see that they had yet to achieve their ideal, they could equally believe that those who came after them would finish the job. The power of reason implied the ineluctability of progress. Science would become more grandiose, more encompassing in scope, yet at the same time more detailed, more scrupulous. Nature was knowable—and if it was knowable then one day, necessarily, it would be known.

This classical vision, springing from the physical sciences, became in the nineteenth century the dominant model for science of all kinds. Geologists, biologists, even the first generation of psychologists, pictured the natural world in its entirety as an intricate but inerrant machine. All sciences aspired to the ideal that physics offered. The trick was to define your science in terms of observations and phenomena that lent themselves to precise description—reducible to numbers, that is—and then to find mathematical laws that tied those numbers into an inescapable system.

No doubt the task was hard. If ever scientists were daunted by their ambitions, it was because of the sheer complexity of the machine they were trying to tease apart. Perhaps the laws of nature would be too vast for their brains to fathom. Perhaps scientists would find they could write down the laws of nature only to discover they lacked the analytical and calculational firepower to work out the consequences. If the project of absolute scientific comprehension were to falter, it would be because the human mind wasn't up to the task, not because nature itself was intractable.

And that's why Heisenberg's argument proved so unsettling. It targeted an unsuspected weakness in the edifice of science—in the substructure, so to speak, a part of the foundation that had gone unexamined because it had seemed so self-evidently secure.

Heisenberg took no issue with the perfectibility of the laws of nature. Instead, it was in the very *facts* of nature that he found strange and alarming difficulties. His uncertainty principle concerned the most elementary act of science: How do we acquire knowledge about the world, the kind of knowledge that we can subject to scientific scrutiny? How, in the particular example Heisenberg took, do we know where some object is and how fast it is moving? It was a question that would have baffled Heisenberg's predecessors. At any time, a moving object has some speed and position. There are ways of measuring or observing these things. The better your observation, the more accurate the result. What else is there to say?

Plenty more, Heisenberg discovered. His conclusion, so revolutionary and esoteric, has been expressed in words that have become almost commonplace. You can measure the speed of a particle, or you can measure its position, but you can't measure both. Or: the more precisely you find out the position, the less well you can know its speed. Or, more indirectly and less obviously: the act of observing changes the thing observed.

The bottom line, at any rate, seems to be that facts are not the simple, hard things they were supposed to be. In the classical picture of the natural world as a great machine, it had been taken for granted that all the working parts of the machinery could be defined with limitless precision and that all their interconnections could be exactly understood. Everything had its place, and there was a place for everything. This had seemed both fundamental and essential. To have a hope of comprehending the universe, you had first to assume that you could find out, piece by piece, what all the components of the universe were and what they were doing. Heisenberg, it seemed, was saying that you couldn't always find out what you wanted to know, that your ability even to describe the natural world was circumscribed. If you couldn't describe it as you wished, how could you hope to reason out its laws?

The implications of Heisenberg's discovery were obscure. And it came on the heels of an equally remarkable, equally perplexing insight that Heisenberg had delivered just two years earlier, when in a visionary flash he saw how to build the theory that became known as quantum mechanics. While the rest of the physics world struggled to keep up, Heisenberg, with a young man's purity of vision, was eager to forge ahead, rewriting the fundamental rules of physics in an abstruse new theoretical language that even he could not yet claim he fully grasped. But Niels Bohr, a man given to slow and sometimes exasperatingly careful reflection, saw the need to assimilate the new to the old. The difficult but essential task, he saw, was to make sense of the new quantum physics without throwing overboard the hard-won successes of the previous era. He and Heisenberg wrangled painfully over how best to portray the emerging, still controversial science.

Another voice came into the argument. By the time Heisenberg announced his principle, Albert Einstein was close to fifty. He was the old man of science, respected, revered, but no longer always attended to. Younger scientists were doing the important work. Einstein occupied the role of lofty commentator. He too, in his day, had been a revolutionary. In his great year of 1905, with his theory of relativity, he had overthrown the old Newtonian idea of absolute space and time. Events that one observer saw as simultaneous might seem to another to happen in sequence, one after the other. A third observer might see that sequence reversed. Heisenberg loosely adduced Einstein's revolutionary principle in support of his own: different observers see the world differently.

But this, to Einstein, was a monstrous misrepresentation of his own greatest achievement. Relativity, to be sure, allowed for differing perspectives, but the whole point of his theory was that it allowed apparently contradictory observations to be reconciled in a way that all observers could accept. In Heisenberg's world, as far as Einstein could see, the very idea of a true fact seemed to crumble into an assortment of irreconcilable points of view. And that, said Einstein, was unacceptable, if science was to mean anything reliable. Here was another fierce intellectual struggle, Heisenberg and Bohr this time joining arms against the old master.

From this shifting, three-way debate there eventually emerged a practical, workaday definition of the uncertainty principle that most physicists continue to find convenient and at least moderately comprehensible—as long as they choose not to think too hard about the still unresolved philosophical or metaphysical difficulties it throws up. Reluctantly, Einstein conceded the technical correctness of the system Heisenberg and Bohr laid out. But he could never accept that it was the last word. To him, the new physics remained until his dying day an unsatisfactory compromise, an interim measure that must eventually be supplanted by a theory resting on the old principles he cherished. Heisenberg's uncertainty, Einstein stoutly insisted, was a sign of human inability to comprehend the physical world, not an indication of something strange and inaccessible about the world itself.

Einstein's profound distaste for the kind of physics that Bohr and Heisenberg were forging blossomed into what was indeed a struggle for the soul of science. Now that the battle is over, that phrase may seem melodramatic. But in the 1920s, when this new physics was emerging, it was all too evident that the foundations of physical science had come under an unprecedented scrutiny. And cracks showed. With Bohr overseeing the task, the foundations were rebuilt—or, as Einstein might have said, propped up—while the superstructure remained more or less as it was. This remarkable rehabilitation forms the core of the story this book tells. Among the principals there were no neutral voices. Nor was it a matter of one side being clearly delineated against another. Allegiances shifted. Views changed. And even now, Einstein's skeptical spirit lingers over the ostensible victory claimed by Bohr and his adherents.

This central story has both an afterword and a preface.

The uncertainty principle has become a catchphrase for the general difficulty, not just in science, of establishing untainted knowledge. When journalists admit that their own views can influence the stories they are reporting, or when anthropologists lament how their presence disrupts the behavior of the cultures they are examining, Heisenberg's principle is not far away: *The observer changes the thing observed.* When literary theorists assert that a text offers a variety of meanings, according to the tastes and prejudices of different readers, Heisenberg lurks in the background: *The act of observation determines what is and isn't observed.*

Does this have anything to do with basic physics? Hardly! Why, then, has Heisenberg's principle been so enthusiastically appropriated by other disciplines? This curious annexation of an esoteric idea arises, I suggest later, not so much because journalists, anthropologists, literary critics, and the like are eager to find dubious scientific justification for their own assertions, but rather because the uncertainty principle makes scientific knowledge itself less daunting to the nonscientists and more like the slippery, elusive kind of knowing we daily grapple with.

To get to that part of the story, however, we must first understand where Heisenberg's uncertainty came from. Scientific revolutions, like any other kind, do not arrive out of thin air. They have roots and antecedents. Uncertainty represents the culmination of quantum mechanics, which by 1927 had already overturned many of the old convictions of classical, nineteenth-century physics. But quantum mechanics was itself a response to problems that the older physics could not handle. Certainty, in science, has always been a fraught issue, and although quantum theory and Heisenberg's uncertainty are unquestionably products of the twentieth century, their earliest glimmerings appeared almost one hundred years earlier. So it is that the tale begins in the opening decades of the nineteenth century.

Chapter 1

IRRITABLE PARTICLES

Robert Brown, son of a Scottish clergyman, was the archetypal self-made scholar, sober, diligent, and careful to the point of fanaticism. Born in 1773, he trained in medicine at Edinburgh, then served for some years as a surgeon's assistant in a Fifeshire regiment. There he put his spare time to worthy use. Rising early, he taught himself German (nouns and their declensions before breakfast, his diary records, conjugation of auxiliary verbs afterward) so that he could master the considerable German literature on botany, his chosen subject. On a visit to London in 1798, the young Scotsman met and so impressed the great botanist Sir Joseph Banks, president of the Royal Society, that on Banks's recommendation he sailed three years later on a long voyage to Australia, returning in 1805 with close to four thousand exotic plant specimens neatly stowed on his ship. These he spent the next several years assiduously describing, classifying, and cataloging, serving meanwhile as Banks's librarian and personal assistant. Brown's remarkable Australian trove, along with Banks's own equally notable collection, became the heart of the botanical department of the British Museum, of which Brown became the first professional curator. He was, said a visitor to Banks's London house, "a *walking catalogue* of every book in the world."

Charles Darwin, before he was married, passed many a Sunday with the learned Robert Brown. In his autobiography Darwin describes a contradictory man, vastly knowledgeable but powerfully inclined to pedantry, generous in some ways, crabbed and suspicious in others. "He seemed to me to be chiefly remarkable for the minuteness of his observations and their perfect accuracy. He never propounded to me any large scientific views in biology," Darwin writes. "He poured out his knowledge to me in the most unreserved manner, yet was strangely jealous on some points." Brown was notorious, Darwin adds, for refusing to lend out specimens from his vast collection, even specimens that no one else possessed and which he knew he would never make any use of himself.

It is ironic, then, that this dry, cautious man should be commemorated now mainly as the observer of a curious phenomenon, Brownian motion, that represented the capricious intrusion of randomness and unpredictability into the elegant mansion of Victorian science. It was indeed the very scrupulousness of Brown's observations that made the implications of Brownian motion so grave.

In June 1827, Brown began a study of pollen grains from *Clarkia pulchella*, a wildflower, popular today with gardeners, that had been discovered in Idaho in 1806 by Meriwether Lewis but named by him for his co-explorer William Clark. Characteristically, he intended to scrutinize minutely the shape and size of

pollen particles, hoping that this would shed light on their function and on the way they interacted with other parts of the plant
to fulfill their reproductive role.

Brown had acquired a microscope of recent and improved design. Its compound lenses largely banished the rainbow-hued
fringes of color that afflicted the borders of objects seen in more
primitive instruments. Under Brown's eye the ghostly shapes of
the pollen grains sprang clearly into view, their edges neatly delineated. Even so, the images were not perfect. The pollen grains
wouldn't stay still. They moved about, jiggled endlessly this way
and that; they shimmered and stuttered; they drifted with strange
erratic grace across the microscope's field of view.

This incessant motion complicated Brown's planned investi-
gations, but it was not so very surprising. More than a century
and a half earlier Antony van Leeuwenhoek, a draper from Delft,
Holland, had astonished and delighted the scientific world when
he described tiny "animalcules" of strange and myriad form that
his crude microscope revealed in droplets of pond water, in
scrapings from the unbrushed teeth of old men, and even in a
suspension of ordinary household pepper crushed into plain water. "The motion of most of these animalcules in the water was
so swift, and so various, upwards, downwards, and round about,
that 'twas wonderful to see," the entranced Leeuwenhoek wrote.
His discovery not only spurred further scientific investigation but
also led well-to-do citizens to purchase microscopes for their parlors and drawing rooms, where they could amaze their guests
with this new wonder of nature.

Some animalcules had tiny hairs or finny extensions that en-
abled them to swim about. Others wriggled like little eels. It was
easy to imagine that their meanderings were purposeful in some
rudimentary way. Pollen grains, on the other hand, were simple
in shape and had no moving parts. Still, they were undeniably

organic. It seemed to Brown not unreasonable to suppose that pollen grains—especially as they were the male parts of a plant's reproductive equipment—might possess some vital spirit that impelled them to move in their amusing but inscrutable fashion.

But Brown distrusted vague hypotheses of this sort. Observation, not speculation, was his forte. Testing pollen from other plants, he found that those grains danced about too. But then he examined minute fragments of leaves and stems and saw that those also jogged perplexingly about under his microscope's gaze.

His attention caught by this "very unexpected fact of seeming vitality," Brown could not help but probe the matter further. He obtained dust from dried plant samples, some more than a century old. He scraped tiny fragments from a piece of petrified wood. All these tiny grains had once been living but were now long dead, any vital spark extinguished. Examined under a microscope, they also shimmied. He moved on to truly inorganic material, knocking tiny shards from a variety of rocks and a piece of ordinary window glass. They too jiggled about. To put the case to the ultimate test, he scratched powder from a piece of the Sphinx, to which, as a curator of the British Museum, he had easy access and which he presumably regarded as certifiably, unarguably dead, on account of its provenance.

Placed in a drop of water under his microscope lens, ancient dust from the Sphinx danced about like everything else.

Brown acknowledged that he was not the first to see things jiggling about under the microscope. A certain Mr. Bywater of Liverpool, he noted, had a few years earlier looked at fragments of both organic and inorganic materials and observed "the animated and irritable particles" that they all shed. But Brown, through a variety of ingenious and careful experiments, established that the ceaseless motion of all these tiny fragments was neither the "animalcular motion" Leeuwenhoek and others had seen nor movement produced by vibration or turbulence of the fluid suspensions, by the action of heat, or through electric or magnetic influences.

This was contradictory and baffling. Dead particles of dust clearly couldn't move of their own volition, nor was any external influence pushing them around. Yet move they all too plainly did. Brown himself made no attempt at an explanation. He was a careful descriptive botanist, not a philosopher of nature, and as Charles Darwin said, "much died with him, owing to his excessive fear of never making a mistake."

Faced with this impossible dilemma, science took the prudent course and ignored Brownian motion for decades. Its deep significance went unnoticed because the phenomenon was so far beyond scientific apprehension. There was no way even to begin to grasp what it meant. Anyone who used a microscope knew about Brownian motion, at least as a great nuisance, but few read carefully what Brown himself had said about it. Most botanists and zoologists persisted in the idea that it was a manifestation of vital spirits, conveniently ignoring Brown's demonstration that inert particles jiggled about just as much. Or else they decided their specimens were buffeted by heat or vibrations or electrical disturbances, ignoring Brown's experiments that ruled out those and other influences.

It was not until after Brown's death in 1858 that a few scientists began to see their way to an understanding of the phenomenon. As often happens in science, the observations could not be understood until there was at least the glimmering of a theory by which to understand them. The theory in this case was not a new idea but a very old one that science finally had the means to make sense of.

The Greek thinker Democritus, who flourished around 400 B.C, believed that all matter was made of tiny, fundamental particles called atoms (from the Greek *atomos*, indivisible). No matter how prescient this notion seems in retrospect, it was really a philosophical conceit more than a scientific hypothesis. What atoms were, what they looked like, how they behaved, how they interacted—such things could only be guessed at. Modern interest in atoms revived first among the chemists. In 1803, John Dalton in England proposed that rules of proportion in chemical reactions—hydrogen and oxygen combining in a fixed ratio to produce water, for example—came about because atoms of chemical substances joined together according to simple numerical rules.

Atoms didn't gain credibility overnight. As late as 1860 an international conference convened in Karlsruhe, Germany, to debate the atomic hypothesis. The weight of opinion by now favored atoms, but there was significant dissent. Many distinguished chemists were happy to take laws of chemical combination as basic rules in their own right, and saw no reason to indulge in extravagant speculation about invisible particles.

August Kekulé, the German chemist who famously devised the ring structure of the benzene molecule while dozing in his fireside armchair and dreaming about snakes catching their own tails, offered a more shaded opinion. He accepted the existence of the chemical atom, along the lines suggested by Dalton and others, and he noted that some physicists, for their own reasons, had also recently begun arguing for atoms. But were the chemist's atom and the physicist's atom the same thing? Kekulé thought not, or at least that any such judgment was premature.

To the chemist, an atom was a thing of almost tactile qualities. It possessed in some way the characteristics of the substance it represented, and it could hook up with or detach from other atoms, according to their respective qualities. Chemists mostly imagined that atoms in bulk sat still, filling space like oranges in a crate.

Physicists thought quite differently. Their atoms were tiny, hard pellets, flying about at high speed in mostly empty space, occasionally crashing into each other and bouncing off again. The role of these atoms was specific. Beginning about halfway through the nineteenth century, a number of mathematically inclined physicists began to pursue the idea that the frenetic motion of atoms could explain the hitherto mysterious phenomenon of heat. As atoms in a volume of gas gained energy, they would fly about faster, bump into each other more violently, and crash more forcefully into the walls of a container. This was why gas would expand when heated, and exert more pressure. In this so-called kinetic theory of heat, heat was nothing but the energy of atomic motion. The deeper implication was that large-scale physics of heat and gases ought to follow ineluctably from the small-scale behavior of atoms as they moved and collided in strict obedience to Newton's laws of motion.

Thus arose the reliable cliché of the atom as a tiny billiard ball, hard but inert, banging mindlessly about. Whether this had anything to do with chemistry was another question. Physicists allowed that the atoms of one gas might be lighter or heavier than those of another, but why gases had distinct chemical properties was none of their business.

The atom, in short, was in these early days by no means a unifying hypothesis. If chemists and physicists had little to say to each other, still further excluded were the microscopists and biologists. Kinetic theory came with mathematical complexities that repelled all but a select few, while the typical mathematician, if even aware of Brownian motion, most likely assumed it was a trivial phenomenon of strictly botanical interest.

Nevertheless, a connection awaited discovery. A first hint came from Ludwig Christian Wiener, who spent most of his life teaching mathematics and geometry at German universities. In

1863, after conducting experiments that confirmed everything Brown had long ago found, Wiener felt able to publish an intriguing if speculative suggestion. If the liquid in which Brownian particles jiggle about was in reality a welter of furious atoms, then these atoms would buffet the suspended particles from all sides. The erratic but incessant agitation of the invisible atoms, he argued, would cause the larger visible particles to stagger unpredictably about.

In keeping with the tangled history of this subject, Wiener's daring proposal attracted next to no interest.

It fell to a series of French and Belgian Jesuit priests to keep digging for a scientific account of Brownian motion. During the nineteenth century many clergymen maintained an active and useful interest in the observational and collecting sciences: botany, geology, zoology, and so on. The clerical connection comes up in *Middlemarch*, when the determinedly atheistic man of science Dr. Lydgate visits the agreeably untheological Reverend Farebrother and finds that the clergyman has an impressive natural history collection, including specimens, books, and journals. Happy to encounter a fellow philosopher of nature, Lydgate offers to show Farebrother a few of his own items, in particular "Brown's new thing — *Microscopic Observations on the Pollen of Plants* — if you don't happen to have it already."

What's more, Jesuits and many other churchmen had a surprisingly broad and rigorous education in philosophy, logic, and even mathematics. Such men were singularly well equipped to deal with problems that we might now call cross-disciplinary but which in those days were merely part of the broad enterprise called science. Mathematical physicists, by contrast, were by the latter half of the nineteenth century on their way to becoming a breed apart, inhabitants of their own recondite discipline, a realm that even those with an adequate amateur command of mathematics were increasingly shy of entering.

This growing divide meant that by the end of the 1870s, a number of scientists perceived the correct qualitative explanation of Brownian motion but lacked the means to put their hypothesis into convincingly quantitative terms. It's strangely difficult to find anyone willing to take credit for seeing the answer. In an 1877 issue of the London *Monthly Microscopical Journal*, for example, we find Father Joseph Delsaulx, S.J., attributing to an unnamed colleague the suggestion that Brownian motion results from the constant agitation of small particles by the atoms or molecules that make up a liquid. (Chemists had by this time established the distinction between atoms, which were fundamental, and molecules, which were combinations of atoms.)

Three years later, writing for the *Revue des Questions Scientifiques*, Father J. Thirion, S.J., mentions that he had seen some years before a similar proposal jotted down in the unpublished laboratory notes (!) of "Fr. Carbonelle, a *savant* well known to our readers, to whom another of our colleagues, Fr. Renard, had shown for the first time the curious movement of *libelles*." These *libelles*, Thirion is helpful enough to explain, are microscopic dark spots seen within little pockets of liquid trapped in samples of quartz. They are in fact tiny bubbles of gas caught within these liquid inclusions, and they jiggle about in the now-familiar fashion. Father Delsaulx also refers to *libelles*, and adds that since quartz is known to be very old, this is a case of Brownian motion that must have been going on unabated for millions of years. Clearly, he says, no external cause can be responsible. What Father Renard showed to Father Carbonelle must be the result, Father Delsaulx affirms, of molecules bouncing endlessly around.

The reverend gentlemen were on the right track, but their

lack of mathematical sophistication prevented them from going much further. Delsaulx suggested vaguely that the observed amplitude of Brownian motion—how far and fast a particle travels on each zig or zag—must have something to do with what he called the "law of large numbers." A molecule of liquid, it was clear by this time, is far too small for a single impact with a Brownian particle to cause any observable motion. Rather, the molecules crash constantly into the particle from all sides, but not quite uniformly. Variations in the impacts on different sides of the Brownian particles make it jiggle about; at the same time, the greater the number of molecules involved, the more their random impacts would tend to cancel each other out, making the motion smaller. The "law of large numbers," by which Delsaulx evidently means to imply some sort of statistical reasoning, should in principle connect the magnitude of Brownian motion with the size and number and speed of molecules in the liquid. More than this he could not say.

Ten years later a nonclerical French scientist, Louis-Georges Gouy, performed a series of careful experiments on Brownian motion, which he nicely described as "*une trépidation constante et caractéristique*." He commented that even now, sixty years after Brown's decisive work, the general opinion seemed to be that the motion was "an accident produced by some external agitation." But as he goes on to say (repeating what Brown and Wiener and various Jesuits had already said), this was clearly not the case. His experiments established—yet again—that the motion occurs for all kinds of particles in any kind of liquid. He could not find any kind of particle that did *not* jiggle about. He concluded, as many had done by now, that molecular activity was the cause.

But he ventured a little further. He first assured his readers that Brownian motion was not, as some had suggested, a kind of perpetual motion, which the recently formulated laws of thermo-dynamics clearly forbade. As molecules bounce around, he explained, they crash into each other, exchanging energy, some going a little slower, some a little faster—but always sharing the same total amount of energy. No problem there. Then he noted that recent estimates put the typical speed of molecules in the region of 100 million times greater than the speed at which Brownian particles appeared to move. Again, no problem: the "law of large numbers" would take care of that. But like Father Delsaulx, he could offer no specific calculation relating the size of a particle, the number of molecules in it, or the number of times it got hit by liquid molecules to the way it moved.

Brownian motion, evidently, was a *statistical* phenomenon: the unpredictable, apparently random jiggling of tiny particles reflected in some way the average or aggregate motion of unseen molecules. It might not be possible to explain in minute detail just why a Brownian particle moved as it did, in its erratic way, but the broad parameters of its movement ought to follow from some suitable statistical measure of the motion of unseen molecules.

But the few early investigators who perceived this connection lacked the means to make their theorizing mathematically precise. And perhaps because they could offer no specific calculation, they failed to see the conceptual puzzle that arose. If the underlying motion of molecules followed orderly Newtonian rules of cause and effect, of perfect predictability, how could it give rise to a phenomenon that appeared to demonstrate the workings of chance? That conundrum, however, was precisely what the more sophisticated advocates of kinetic theory soon found themselves obliged to contend with.

Chapter 2

ENTROPY STRIVES TOWARD A MAXIMUM

Writing about Brownian motion in 1889, Louis-Georges Gouy expressed some perplexity that "this phenomenon seems to have barely attracted the attention of physicists." Among those who had failed to grasp its significance, he claimed, was the Scottish physicist James Clerk Maxwell, arguably the most eminent theorist of the nineteenth century, who had apparently believed that if "submitted to a more powerful microscope . . . the [Brownian particles] will demonstrate only more perfect repose." Better optics, in other words, would make this nuisance go away.

Unfortunately Gouy, as was often the case in those days, gave no source for Maxwell's alleged statement, and it remains unclear even now whether his accusation is just. Certainly, though, nothing in Maxwell's published works hints that he saw in

Brownian motion any clue to the molecular composition of gases and liquids. His omission is all the more surprising since it was Maxwell who first used a statistical technique to solve a problem in physics and who later helped develop the kinetic theory of heat to a fine mathematical pitch.

As far back as the middle of the seventeenth century, Blaise Pascal, Pierre de Fermat, and others had worked out simple laws of mathematical probability for various card and dice games, but it was a long time before such ideas left the gambling parlors. In 1831, the Belgian mathematician Adolphe Quetelet tabulated crime rates in France according to the age, sex, and education of the perpetrator, the climate at the crime location, and the time of year when the crime occurred. This heralded, for good or ill, the widespread application of statistical methods in demographics and the social sciences.

About thirty years later Maxwell, influenced by his reading of Quetelet, came up with an ingenious way to prove that the rings of Saturn must be made of dust. Picturing the rings as aggregations of small particles controlled by Saturn's gravity, Maxwell set up a statistical description that allowed the particles to have a range of sizes and orbital speeds. Applying standard mechanical analysis to this model, he proved that if the rings were to maintain their shape over long periods of time, the particle sizes must fall within some limited range.

Shortly afterward, Maxwell saw that he could use similar methods to describe the speeding and colliding atoms that constituted a volume of gas. It was in understanding the nature of heat that physicists first found themselves obliged to tackle seriously questions of statistics and probability. But from the outset, there was something disquieting, almost self-contradictory, about this venture.

If heat is simply the collective bustle of atoms, then the

physics of heat must ultimately follow from Newton's laws of motion applied to those atoms. Atomic collisions ought to be as calculable as caroms on the billiard table, in which case the behavior of heat ought to be similarly predictable. This vision of scientific omniscience, in which every particle in the universe must follow strict and rational laws, is captured in the famous words of the Marquis de Laplace, one of the leading eighteenth-century developers of Newtonianism at its mathematically splendid best:

> We may regard the present state of the universe as the effect of its past and the cause of its future. An intellect which at any given moment knew all of the forces that animate nature and the mutual positions of the beings that compose it, if this intellect were vast enough to submit the data to analysis, could condense into a single formula the movement of the greatest bodies of the universe and that of the lightest atom; for such an intellect nothing could be uncertain and the future just like the past would be present before its eyes.

Nothing could be uncertain: that was the crucial point. Mangling the words of another Frenchman, we can say: *Tout comprendre c'est tout prédire*, to understand all is to predict all. From such grandiose contentions derive all the familiar clichés about the world as machine, the universe as clockwork, and all of science as ultimately deterministic and inexorable.

On the other hand, as physicists quickly realized, any hope of actually calculating the individual behavior of every last atom or molecule in a volume of gas was not just unattainable but positively absurd. (By the latter half of the nineteenth century, scientists had a pretty good idea of how tiny and therefore how numerous molecules must be. A laboratory flask filled with water contains a trillion trillion of them.) In order to accomplish anything practical by way of theorizing about large crowds of atoms and molecules, physicists would have to resort to statistical descriptions of their behavior and set aside the utopian goal of perfect knowledge. Nowhere did this compromise show up more disturbingly than in the notorious second law of thermodynamics—the one about entropy, and the contest between order and disorder.

Heat flows from hot bodies to cold ones but not, evidently, the other way around. *Entropie strebt ein Maximum zu*, the German physicist Rudolf Clausius declared in 1865, coining a new word: "Entropy strives toward a maximum." That maximum is achieved when heat spreads itself around as evenly and uniformly as possible. Put a couple of ice cubes in a summer drink. Heat flows from the liquid to the cold ice, so that the ice melts and the drink becomes cooler. In the process, entropy increases. If the ice cubes grew bigger while the cold drink around them began to boil, entropy would be decreasing, and that's what the second law forbids.

Clausius and others propounded the second law before the nature of heat was truly understood. They took the law to be rigid and exact, as laws of physics are supposed to be. Heat *always* flows from hot to cold. Entropy can *only* increase.

The realization that heat is nothing but the motion of atoms seemed at first to clarify the second law. If a collection of fast-moving, therefore hot, atoms is mixed with a collection of slow-moving, therefore cold, atoms, it's not hard to understand that random bashing about among the atoms would tend to slow down the fast ones and buck up the slow ones until all, on average, are moving at the same speed. The temperature would then be the same everywhere, and entropy would have been duly maximized. The opposite process—the fast atoms getting faster, taking energy from the slower ones, which become slower still—doesn't seem plausible.

In 1877, the prickly, irascible Austrian physicist Ludwig Boltzmann proved a difficult mathematical theorem saying exactly this. He found a way to define entropy as a statistical measure of the motion of a collection of atoms, and showed that collisions among atoms would push entropy toward its maximum value. It's from Boltzmann that we get the idea of entropy having to do with order or disorder. If, in some container of gas, all the fast atoms hung about one end while all the slow ones stayed at the other, that would be a state with an unusual degree of arrangement or orderliness. It would have low entropy. Let all the atoms mix, collide, and share their energy as equably as possible, and they attain a state of maximum entropy. The atoms are then at maximum disorder, in the sense that they are as randomly arranged as possible. Ignorance of what the atoms are up to is spread uniformly.

But something about Boltzmann's theorem didn't seem right. The increase of entropy represents directionality—a process that always goes one way, never the other. Yet Newton's laws, governing the movements of atoms, are thoroughly evenhanded with respect to time. A set of atomic motions, if played in a time-reversed manner, will still obey Newton's laws. Mechanics contains no intrinsic distinction between past and future, whereas in Boltzmann's theorem, elaborately derived from mechanics, that distinction mysteriously appears.

Not too many years after Boltzmann had proved his theorem, the French mathematician Henri Poincaré proved a theorem of his own that seemed to contradict Boltzmann. Applied to a set of atoms constituting a gas, Poincaré's theorem said that every possible arrangement of atoms, corresponding to states with entropy high, low, and in between, must occur sooner or later, in the fullness of time. In which case, it would seem, entropy can and must decrease as well as increase.

Perplexities such as these led some physicists to an extreme viewpoint: atoms cannot be real, they said, since they lead to theoretical paradox. In some quarters this conclusion was warmly received. In the German-speaking world especially there had arisen a so-called positivist philosophy of science, whose adherents argued that atoms were illegitimate in the first place. Science, they said, should deal in what is visible and tangible, in what experimenters can directly observe and measure. That means atoms are at best speculative, and reasoning based on them is strictly hypothetical. Atoms are not, the positivists sternly maintained, the factual, trustworthy ingredients from which real science should be made.

Tortuous attempts to resolve the apparent contradiction between Boltzmann's and Poincaré's theorems only made the positivists happier. The gist of it is that Boltzmann's theorem—because of certain assumptions he had been obliged to make in order to untangle the fearsome mathematics he had gotten himself into—is not exactly true. Generally, it's far more likely that orderly arrangements of atoms will become disorderly than vice versa—but the latter is not completely ruled out.

With this qualification, physicists realized that their kinetic theory of heat was telling them something rather unexpected and subtle. It is not absolutely certain, they saw, that entropy must always increase, that heat must always flow from hot to cold. There is a chance, depending on the way atoms happen to crash around, that a bit of heat could move from a cool place to a hotter one, so that entropy would for a moment decrease. Probability comes irrevocably into the picture. Most of the time, everything will happen in the expected way. Collisions among atoms will almost always tend to increase disorder, and therefore increase entropy. But the reverse is not impossible, only unlikely.

This dubious and slippery conclusion outraged the positivists further. If a law of physics were to mean anything, they said, it must surely be definitive. To say that heat would most likely flow from hot to cold, but that it had some chance, no matter how tiny, of going the other way, was to make a mockery of scientific thinking. Still further reason to disbelieve in the fiction called atoms.

It became an urgent matter for the pro-atom physicists to bolster their case in a way that positivists would accept. In 1896 Boltzmann himself, in a reply to one of his critics, hit upon a straightforward and easy argument in favor of atoms. "The observed motions of very small particles in a gas," he wrote, "may be due to the circumstance that the pressure exerted on their surfaces by the gas is sometimes a little greater, sometimes a little smaller." In other words, because a gas is made of atoms, and because these atoms dance around in an erratic way, a small particle within the gas will be jostled unpredictably back and forth. This is precisely what Gouy, following Fathers Thirion and Delsaulx, had already said, but Boltzmann evidently knew nothing of their work. He was the first physicist of profound mathematical ability to hit upon the idea that Brownian motion provides direct visual evidence not only of the atomic nature of matter but of the randomness inherent in atomic motions.

This throwaway remark by Boltzmann caught no one's attention, and indeed has barely been noticed by historians of science ever since. His casual manner suggests that he didn't find the suggestion either novel or particularly important. Like Thirion, Delsaulx, and Gouy, he took it as unremarkable that molecular movement explained Brownian motion. Unlike those previous authors, who made vague references to the "law of large numbers," Boltzmann had sufficient mastery of statistical theory that he could have attempted to calculate the expected magnitude of Brownian motion in terms of the underlying movements of atoms.

But he didn't make the effort. Maxwell had failed to heed what Brownian motion was telling physicists. Now Boltzmann got the message but, perhaps thinking the point perfectly obvious, didn't pursue it.

Another decade passed before the tale of Brownian motion reached its momentous conclusion, and it is at this point in our story that we first encounter the piercing intellect of Albert Einstein. In 1905 Einstein was a smart, trim twenty-six-year-old, working at the patent office in Bern because he had been unable to land an academic position. He had published a few papers but was far from well known in the physics community. That was about to change.

An admirer of Boltzmann's dense and, frankly, long-winded monographs, Einstein had become fascinated by statistical questions in physics and by the attendant controversy over the existence of atoms. At some point he too realized that a suitably small particle immersed in a liquid would bounce about because of molecular collisions—exactly as Boltzmann had said, though it appears that Einstein, like everyone else, failed to notice this obscure remark of his predecessor. In any case, Einstein dug deeper. He wondered if the motion of a particle large enough to see in a microscope might constitute a direct and quantitative test of the atomic hypothesis—exactly what the positivists demanded but said was impossible. And so he decided to calculate the answer.

It was not a straightforward piece of reasoning. Gouy had realized that a Brownian particle ought to have, on average, the same energy of motion as the molecules of the liquid in which it was suspended. Those molecules, being many times less massive, would zip around very fast, while the Brownian particle would blunder about much more slowly. There should be a simple relationship between the average speed of the molecule and the average speed of the particle in the liquid. But the erratic nature of Brownian motion made it hard to define a particle's average speed in a meaningful way, and an experimenter of the late nineteenth century was in no position to measure or record precisely how such a particle zigged and zagged.

Ingeniously, Einstein took a different tack. He found a way to calculate not how fast a suspended particle would move but how far its hither-and-thither motions would cause it to drift in some period of time. For example, one could draw a small circle around the starting position of some particle and ask how long it should take, on average, to reach the circumference. In this way Einstein derived a theoretical result that could be put to practical scrutiny. Finally, almost eighty years after Robert Brown had given a scientific account of the motion of small particles suspended in fluid, Einstein provided the first quantitative treatment of its true cause. His clever analysis constituted one of the four historic papers he published in his annus mirabilis of 1905; in the others he propounded his special theory of relativity to what was then a mainly bewildered audience of physicists, and offered provocative notions about the true nature of light.

In a final exasperating irony, it turns out that Einstein, when he began his calculations, didn't even know there was such a thing as Brownian motion. Only in the course of writing his paper did he discover that the phenomenon had been known to microscopists, botanists, and others for generations. In his introduction he cautioned that "it is possible that the motions to be discussed here are identical with the so-called 'Brownian molecular motion'; however, the details I have been able to ascertain regarding the latter are so imprecise that I can form no judgment in the matter."

Three years later, in 1908, the French physicist Jean Perrin performed a series of careful experiments to measure Brownian motion and compare the findings with Einstein's theory. It all matched up, and Perrin's work is often cited as the crucial, crushing evidence for the existence of atoms. For most physicists this came as no surprise but rather was pleasant confirmation of what they had long believed. Even the most die-hard positivist opponents of atomism, with one or two exceptions, now had to give way.

From this point on, atoms were undeniably real. At the same time statistical thinking was cemented firmly into place as an essential part of physical theorizing. The two were inextricably tied together. Those who had been espousing kinetic theory for years took satisfaction in this development: any useful account of atoms necessarily involves statistical reasoning. The chancy nature of the second law of thermodynamics—entropy almost always rises—was here to stay.

Even so, determinism survived—or seemed to. To Einstein, certainly, the appeal of statistical reasoning was precisely that it allowed the physicist to make quantitative statements about the behavior of crowds of atoms, even while the motion of individual atoms remained beyond the observer's ken. What mattered was only that those motions followed strict and unerring rules. Nature, at bottom, remained intrinsically deterministic. The problem is that scientific observers cannot gather all the information they would need to fulfill Laplace's ideal of total knowledge leading to perfect predictability.

Without altogether appreciating what had happened, physicists had subtly revised their estimation of what a theory meant.

Until this time, a theory was a set of rules that accounted for some set of facts. Between theory and experiment there existed a direct, exact two-way correspondence. But that was no longer quite the case. Theory now contained elements that the physicists were sure existed in reality, but which they couldn't get at experimentally. For the theorist, atoms had definite existence, and had definite positions and speeds. For the experimenter, atoms existed only inferentially, and could be described only statistically. A gap had opened up between what a theory said was the full and correct picture of the physical world and what an experiment could in practice reveal of that world.

What was lost, then, was not the underlying ideal of a deterministic physical world but the Laplacian hope for perfectibility in the scientific accounting of that world. The universe unfolds imperturbably according to its inner design. Scientists could legitimately hope to understand that design fully. What they could no longer attain, it seemed, was complete knowledge of how that design was realized. They could know the blueprint, but not the shape and color of every brick.

One commentator who glimpsed this difficulty was the historian Henry Adams, whose idiosyncratic autobiography, *The Education of Henry Adams*, depicts a man of old-fashioned classical wisdom, a scholar of politics and culture and religion, struggling to stay on his feet in a world increasingly driven by science and technology. It wasn't that he was opposed to science, rather that he found its grandiosity and reach forbidding and more than a little alarming.

Adams heard of the advance of statistical reasoning in physics and found it perplexing in a way that most scientists did not care to think about. Science aimed for completion and perfection, of course, but now, as Adams loftily put it, "the scientific synthesis commonly called Unity was the scientific analysis commonly called Multiplicity." It seemed to him, in his somewhat over-heated way, that kinetic theory was but a step away, philosophically, from chaos and anarchy. What was the meaning of the quest for unity and synthesis in science if the power of prediction would from now on only ever be approximate?

Adams quizzed his scientifically and philosophically minded friends, but, he lamented, "here everybody flatly refused help." Perhaps they couldn't grasp what he was getting at. Adams had a fondness for enigmatic, obscure oratory. To scientists it only seemed that their statistical theories actually gave them a greater grasp of the universe and an increased power of prediction. They understood more now than they had before, and would understand still more in the future. Any losses seemed conceptual, metaphysical, philosophical—and therefore of no scientific account.

Chapter 3

AN ENIGMA, A SUBJECT OF PROFOUND ASTONISHMENT

I t would be more accurate to say, by the first decade of the twentieth century, that science had amassed a surfeit of atoms, all doing distinct jobs and with no clear kinship between them. Of some standing were the chemists' atoms, the indivisible units of matter that participated in reactions and joined together to form molecules. Not quite as venerable were the physicists' kinetic atoms, those prototypical billiard balls that in their random crashing around gave substance to the laws of heat. Between these two atoms, from a theoretical perspective, there was essentially no point of contact. And in 1896 a new task had been piled onto the already overburdened atom.

Henri Becquerel's discovery of radioactivity offers eloquent testimony to the power of serendipity. On the first day of January

1896, a German physicist by the name of Wilhelm Röntgen sent to his colleagues across Europe details of an astonishing observation. To prove his point, he included a photograph of his hand — or rather, an eerie likeness of the bones of his hand, with flesh discernible as a faint halo and with the unmistakable shadow of his wedding ring loosely orbiting the skeletal third finger. This was the world's first X-ray image, and it set off a sensation not only among scientists but also in the newspapers, which raced to print pictures of bones, nails accidentally embedded in limbs, and internal skeletal deformities of one kind or another.

Röntgen's discovery was itself purely accidental. He had noticed a strange glow on a phosphorescent screen near an electrical discharge tube in his laboratory and, investigating further, had seen a bony shadow spring into visibility when he placed his hand between tube and screen. Physicists, it turned out, had been making X-rays for years without knowing it. Once the news was out, labs around the world set out to explore these unseen penetrating rays. It was quickly established that they were a kind of electromagnetic radiation shorter in wavelength than visible light and ultraviolet.

Becquerel, seeing X-ray images at a meeting of the French Academy of Sciences in Paris early in 1896, followed a hunch. He was the son and grandson of distinguished Parisian physicists, all graduates of the École Polytechnique, all members of the Académie Française, and all, one after the other, occupants of the chair of physics at the Musée d'Histoire Naturelle. Henri's son Jean, in due course, would follow the same path. The various Becquerels investigated electricity, chemistry, and sunlight, among other things, but one particular interest had become a family tradition. They all studied fluorescence, the phenomenon by which certain minerals, after exposure to strong sunlight, are then seen to emit a faint luminosity of their own when taken into the dark. Henri's father had established himself as an expert especially on the fluorescence of uranium-bearing minerals.

Hearing about X-rays, Henri Becquerel wondered if this curious new emanation had any connection to the fluorescence he knew so much about. His first experiments seemed to confirm that suspicion. He took a variety of fluorescent minerals, including potassium uranyl sulfate (a special favorite of his father's), placed them on photographic plates wrapped tightly in thick black paper, and set the samples in bright sunlight to activate their fluorescence. Developing the plates after a few hours, he found that the one beneath the uranium-containing mineral had been fogged by some emanation that had penetrated the opaque paper. This rock, he concluded, activated by sunlight, was giving off X-rays.

But then, as luck would have it, Paris became gray and overcast. For days sunlight was not to be had. Becquerel tucked his experiments away in a drawer. At some point, perhaps merely to test the integrity of his wrapped photographic plates, Becquerel took one out of its dark hiding place and developed it. To his everlasting surprise, he found that this plate was fogged too. Even though it hadn't been exposed to sunlight, the uranium mineral had given out some sort of radiation that passed through the thick paper and triggered a reaction in the sensitive chemicals. The emission was neither X-rays nor conventional fluorescence, but something new and strange, intrinsic to the mineral itself. *Les rayons uraniques*, Becquerel called them, which summed up everything he knew.

Reporting his odd discovery to the Academy of Sciences, he got a lukewarm response. X-rays continued to fascinate, and Becquerel's fuzzy splotches could hardly compete with images of broken bones. He shrugged and went back to his lab. Formulating an incorrect hypothesis about Röntgen's serendipitous X-rays, Becquerel had performed a misleading experiment, only to be nudged into a more interesting one because of bad weather. Thus did he stumble across an entirely new scientific phenomenon. But serendipity can go only so far, and at this point Becquerel's inspiration ran out. He didn't know what else he could do, and no one else seemed interested.

Not until the end of the following year did *les rayons uraniques* attract the interest of a young researcher on the lookout for an unexplored area in which to make a mark. This novice was Marie Curie, born Maria Sklodowska to schoolteacher parents in Warsaw. Because Poland at that time was under oppressive Russian rule, Maria and her older sister Bronia hatched a plan to seek their freedom elsewhere. Bronia came to Paris to study medicine, while Maria, whom the French called Marie, went in for physics and mathematics. Even in Paris, which was more hospitable to female students than most European cities, this was a brave choice. Accepted wisdom held that the female intellect, if educable at all, was better suited to the softer medical and biological sciences. But Marie, stubborn and independent, forged her own path. She met and married Pierre Curie, a physicist eight years her senior who could be as ornery as she was. The two embarked determinedly on their own course of research.

Unhindered by Becquerel's conviction that uranium was the crucial ingredient, Marie Curie systematically surveyed all kinds of minerals, both common and rare, to see if they gave off penetrating rays. Gold and copper did nothing. All uranium minerals were active, as Becquerel had concluded, but so was the mineral aeschynite, which contained no uranium. Pitchblende, the main uranium ore, was indeed active, but actually too active—it produced emanations in greater intensity than Marie calculated it should, from its known uranium content. In short, she quickly concluded, there were things besides uranium that gave off *les rayons uraniques*.

Together, the Curies embarked on the excruciatingly difficult and finicky task of teasing out from pitchblende the additional sources of the emanations. A chemical separation that extracted the known element bismuth yielded an active residue. Bismuth itself, they knew, was not active. Therefore some new element, chemically similar to bismuth and tagging along with it, must be the source of the activity. Announcing this result in April 1898, the Curies proposed to name this new element polonium, in honor of Marie's homeland. Later in the year they found evidence for a second element that came out in chemical extractions of barium. This they called radium, in a report that also bestowed a new name, radioactivity, on the phenomenon Becquerel had originally discovered.

What came next was one of the most arduous, backbreaking, and downright hazardous efforts in the annals of science. From a pitchblende mine in Joachimsthal, Czechoslovakia (whose metals were used in German coinage, the name of one of which, the *Thaler*, eventually transmuted into the dollar), the Curies obtained ten tons of the residue left after uranium had been extracted. They got the use of a large, rickety shed with a leaky glass roof, whose windows they had to keep open even in bad weather so as to allow noxious fumes to escape. In scenes worthy of *Macbeth*, Marie Curie stirred and boiled cauldrons of ore residues and solvents, reducing tens of kilograms of dross into grams of precious distillates, then combining the distillates and reducing them further to concentrate the radium. Over the next two years she reported to the Academy of Sciences steady progress in her quest to isolate the new element. As the concentration of radium grew, her tiny samples began to glow from their emissions. She and her husband held these fiercely radioactive sources up to their closed eyes and saw flashes and meteors within their eyeballs.

Not until July 1902, after close to four years of scientific hard labor, was Marie Curie able to announce that from ten tons of residue she had been able to extract an entire tenth of a gram of pure radium. The periodic table of the elements, that great organizing system dreamed up by Dmitri Mendeleyev, was just over thirty years old. The discovery of an addition to the table was thrilling, and radium was a strange addition indeed, a new element with mysterious and possibly alarming powers.

Thanks to Marie Curie's herculean efforts, radium began to gain attention. In his self-conscious role of perplexed observer, Henry Adams marveled at the machines and the science on display at the Great Exposition in Paris in 1900. He strolled about in the company of the American astronomer Samuel Langley, who had measured the total energy output of the sun, including its invisible infrared emissions as well as visible light. "His [Langley's] own rays, with which he had doubled the solar spectrum, were altogether harmless and beneficent," Adams related in his magniloquent way, "but Radium denied its God—or, what was to Langley the same thing, denied the truths of his Science. The force was wholly new." Scientists would have demurred, to be sure, at any suggestion that radioactivity was a new god—but undoubtedly it was a phenomenon outside the reach of the physics of the day.

For their discoveries, the Curies shared the 1903 Nobel Prize in Chemistry with Henri Becquerel. They were the codifiers, the taxonomists of the new phenomenon. But what were these radioactive emanations, and what was the process that released them? Marie Curie's talents were less well suited to address these questions. But prescient remarks by her reveal that she could see the nature of the conundrums ahead. Her scrupulous examination of radioactivity from numerous sources had led her to an inescapable conclusion: the intensity of radioactive emissions depends on the amount of the radioactive element in the source, and on nothing else. Not on the chemical form the element takes, not on the temperature of the sample, not on light or dark, not on any electric or magnetic field. "Radioactivity," she wrote in December 1898, "is an atomic property"—meaning that its intensity depends purely and simply on how many atoms of uranium or polonium or radium a sample contains.

Two years later, in a comprehensive review prepared for an international physics meeting in conjunction with the Great Exposition, the Curies offered an even more pregnant statement: "The spontaneity of the radiation," they said, "is an enigma, a subject of profound astonishment."

Spontaneity: that was the strange, crucial factor, and a distinctly awkward one it was for scientists inculcated in nineteenth-century traditions. If a lump of uranium ore, stony and impassive, sits on a laboratory bench and emits invisible rays, where is the operation of cause and effect? Where is the scientifically essential idea that if something happens, it happens for a reason, because some prior event made it happen? Radioactivity, as far as anyone could tell in 1900, was uncaused, and therefore scientifically uncalled for.

What's more, radioactivity released energy. In 1903, Pierre Curie and a collaborator collected a large enough sample of radium to show that its activity could heat a small sample of water to boiling. A demonstration at the annual meeting of the British Association for the Advancement of Science prompted one observer to wonder whether this was not some form of perpetual motion. Did radioactive energy, in its spontaneous way, pop out of nowhere?

Marie Curie leaned toward the idea that the law of conserva-

tion of energy, fifty years old now, was not the absolute injunc-
tion scientists had assumed. Perhaps an atom could somehow
manufacture energy out of nothing and carry on as before. This
was not easy to accept, but it seemed to the Curies and to many
others the least unacceptable of various unpalatable interpreta-
tions of radioactivity's troubling spontaneity.

The man who chiefly resolved this confusion—and in the process
brought the modern atom into being—came roaring onto the
stage from a childhood on a New Zealand farm. Ernest Rutherford was brilliant, inventive, and ebullient. Supported by a schol-
arship for gifted colonials, he came to Cambridge in 1897 to study
under J. J. Thomson—universally known as J.J.—who was then
head of the Cavendish Laboratory. He arrived at a thrilling, opportune moment. Just a few months earlier, Thomson had memo-
rably proved that the vacuum tube emanations known as cathode
rays were in fact not rays but streams of electrically charged particles. The word *electron* entered the language, and a tiny thing the
electron proved to be, less massive by far than any individual atom.
At the same time, courtesy of the Curies, radioactivity was finally
attracting attention. With these fundamental discoveries whirling
around him, Rutherford quickly dropped his earlier interest in the
technology of wireless signal transmission—in which he was for a
time mentioned in the same breath as Marconi—and turned his
mind to serious physics.

Rutherford and his mentor were antipodean in character as
well as geographical origin. J.J. was decidedly old-school, dry in
manner and rather reserved, while Rutherford, a boisterous colo-
nial fellow and a keen sportsman, plowed into Cambridge life in
happy ignorance of its minute gradations of class and social status.

Rutherford was self-confident and generally unself-conscious, but quite shrewd enough to relish his own bumptiousness. There was no doubting his talents. "I have never had a student with more enthusiasm or ability for original research than Mr. Rutherford," wrote Thomson in a testimonial for his outstanding protégé.

Working quickly, Rutherford demonstrated in 1898 that there were at least two different kinds of radioactive emanations. One could be stopped by a thick sheet of cardboard, while the other had far greater penetrative power. These he called alpha and beta types of radioactivity. The identity of the alphas remained unclear, but the beta particles, it soon emerged, were nothing but fast-moving electrons.

Did atoms, therefore, contain electrons? Perhaps—but that could hardly be the whole story, since electrons were light and electrically charged, while atoms were heavy and neutral. Thomson evolved what became known as the "plum pudding" atom, in which a small number of electrons moved about in some fashion within a roughly spherical blob of some kind of positively charged medium, ether, . . . *something*, at any rate, that could supply mass and neutralize the electron's negative charge. Vague as this model was, Thomson used it a few years later to interpret a number of experimental findings, concluding that a hydrogen atom most likely contained just a single electron.

Rutherford had a wary distrust of theorizing. To him it seemed premature to speculate too far about what might be in an atom when no one yet knew what an atom was. From Cambridge he went for a few years to McGill University in Montreal, Canada, where he assembled a team to look further into the alphas and the betas and into the elements that generated them. There Rutherford roamed energetically about his laboratory, praising, questioning, encouraging, and occasionally berating his colleagues and students.

Enlightenment did not come easily. The Curies had already identified a number of radioactive elements. Rutherford and others found many more. A welter of names sprang up: radium A, radium B, and on to E; thorium A, thorium B, thorium X, and also a radioactive gas dubbed thorium emanation; then actinium A, B, and actinium emanation . . . All were somehow distinct, all somehow related.

Teasing out one putative element from another, identifying the circumstances by which one disappeared and another appeared, Rutherford and his students painstakingly untangled the confusion. The sweeping conclusion came in a paper published in 1902 with Frederick Soddy, an Oxford-trained chemist who had joined the McGill team. What Rutherford and Soddy proposed was the transformation theory of radioactivity, or, more daringly, the transmutation theory. The things transmuted, they claimed, were the atoms themselves—the supposedly indivisible building blocks of the elements. They laid out a system by which the multiplicity of radiums and thoriums and actiniums and their emanations could be understood as the links in a chain of radioactive decay, one element turning into another, that product transforming into yet another, and so on, with each transformation accompanied by a certain kind of radioactive emission.

Alchemy! cried many critics. The inviolable identity of the elements was a bedrock principle that chemists, through long and difficult struggles, had only recently established. Now came Rutherford and Soddy, saying that the elements weren't permanent after all. Marie Curie, among others, found the suggestion unacceptable. Atoms were by their very essence unchangeable, she maintained, so that any theory in which they could transform into each other was not an honest theory of atoms at all.

But the ability of the transformation theory to make sense of a welter of radioactive substances, with the aid of only a few simple rules, quickly convinced the scientific world of its essential correctness. One of those rules, however, concealed a notion more subversive even than the transmutation of elements. Each radioactive element, Rutherford and Soddy noted, had a rate of decay characterized by what came to be known as the half-life. Start with a gram of the element known then as thorium X, for example, then wait about eleven minutes, and you would have half a gram left. After another eleven minutes, a quarter of a gram would remain, then an eighth, and so on, closer and closer to zero but never quite getting there.

This is an exponential decay, a simple enough mathematical rule. But think of the sample as a collection of atoms and its disturbing meaning dawns. In any eleven-minute period, half the atoms disintegrate while the other half do nothing. And who is to say which atoms decay and which do not?

As Marie Curie had observed, what made radioactivity so troubling was its spontaneity. Rutherford and Soddy had now made this unpredictability quantitative. Decay follows an elementary law of probability, so that in any given time, each atom has a certain chance of decaying. But what does it mean for the principle of cause and effect if an atom sits there, minding its own business, then at some apparently unpredictable moment bursts apart? What made it decay? What made it decay at all, that is, and what made it decay at that particular time?

Then again randomness, by the early twentieth century, was not so novel or alarming a concept as it had been just a generation earlier. Physicists by now had digested the use of statistical theorizing about atoms in gases and had reluctantly accepted the intrusion of probability in the not-quite-predictable behavior of entropy. If radioactive decay too followed a law of probability, perhaps the underlying reason was not so different.

Suggestions of that sort cropped up. The atom might have internal components—sub-atoms, one physicist proposed—and those components might bash incessantly around just as atoms in a volume of gas teem about. It might happen that occasionally, through their random motions, a handful of these sub-atoms would bunch closely enough together to somehow trigger an instability of the whole atom. This hardly qualified as a theory, but it made probabilistic decay acceptable for the same reason the second law of thermodynamics was acceptable. The commotion of sub-atoms inside atoms follows rigidly deterministic rules, but the physicist observing from outside has no hope of knowing what all those sub-atoms are up to. So randomness emerges out of ignorance. If you could somehow see inside an atom and discern its components, then in principle you could track their motion and predict when that particular atom would decay.

That was the far-off but comforting hope, at any rate. Most physicists simply postponed the question as one that they were in no position to usefully address. To understand the strange rule of probability that governed the decay of radioactive atoms, you first had to understand how an atom was built and how it worked.

Chapter 4

HOW DOES AN ELECTRON DECIDE?

In September 1911, a young Dane just shy of his twenty-sixth birthday arrived in Cambridge to learn electron physics from J. J. Thomson. Niels Bohr was the son of a professor of physiology at the University of Copenhagen. His family, going back three generations, boasted schoolteachers, university professors, and ministers of the church. Bohr had written a doctoral thesis on the conduction of electricity in metals, assuming that electrons carried the current and that they rattled more or less freely about inside a conductor, rather as atoms of a gas might fly up and down a tube. The model didn't work very well, and Bohr already suspected that something was fundamentally amiss with the idea of treating electrons, in nineteenth-century style, as electrically charged billiard balls.

Bohr in repose had a mournful look about him. His heavy

brows overhung his eyes; his even heavier mouth drooped down at the corners. Thinking hard, his features slack and his arms hanging limply at his sides, Bohr could look, said one physicist, like an idiot. In later years he gained a reputation for speaking in a slow, ponderous, cryptic way that alternately charmed and exasperated his listeners, so it comes as a surprise to learn that soon after arriving in Cambridge, he managed to offend J. J. Thomson by offering some terse criticisms of the great man's book on the conduction of electricity by gases.

Bohr had difficulty with English manners. He left a manuscript for J.J. to look at and, on discovering it undisturbed some days later, decided to bring the matter up directly. This was not the done thing. J.J.'s reaction, when it eventually came, was to suggest that a young man such as Bohr could not possibly know as much about electrons as he did. It didn't help, Bohr concluded, that he was a foreigner. He went to formal dinner at Trinity, J.J.'s college, but weeks passed before anyone spoke to him. J.J. responded to Bohr's unmannerly desire to debate physics by retreating the other way whenever he saw Bohr coming. "Very interesting . . . absolutely useless" was how Bohr later described his brief stay in Cambridge. *Very interesting* became his trademark way of politely closing the conversation when presented with dubious hypotheses or fanciful scientific speculation.

Bohr traveled to Manchester to visit a professor there, a friend of his recently deceased father's. At dinner he met Ernest Rutherford, who had returned from Canada some years earlier to take up a post at Manchester and who happened to know the same man. Some weeks later, Rutherford visited Cambridge, and he and Bohr spoke again. Clearly it was Rutherford at Manchester, not Thomson at Cambridge, who was doing the most important physics in England. Rutherford, moreover, was not English, and

Bohr found him friendly and encouraging. By March 1912, Bohr had managed to transfer himself to Manchester, ostensibly to learn how to do experiments in radioactivity. At this he proved, if not absolutely useless, indifferent at best.

Rutherford had forged ahead with his scrutiny of the atom. Some years before, working with his young colleague Hans Geiger (of Geiger counter fame), he had finally pinned down the identity of radioactive alpha emanations. They were particles much heavier than electrons, carrying two positive units of electric charge. Trapped and allowed to become electrically neutral, alpha particles became indistinguishable in all respects, Rutherford and Geiger found, from atoms of helium. In alpha decay, evidently, a big atom turned into a somewhat smaller one by spitting out a chunk closely resembling the light helium atom.

Of course, no one then knew what an atom was, but it occurred to Rutherford that alpha particles would make good heavy projectiles for shooting at other things, to see what they were made of. He and Geiger, along with a new student, Ernest Marsden, experimented with shooting alphas from a radioactive source toward thin gold foils. Geiger and Marsden sat in the dark for hours, letting their eyes grow sensitive to the tiny flashes of light that erupted when alpha particles smashed into phosphorescent screens surrounding the experiment.

They weren't sure what they expected to happen. Most of the time, the alphas sailed straight through the flimsy gold foil as if it wasn't there. Sometimes they changed direction a little as they passed through, coming out on the far side at a modest angle. What took the experimenters wholly by surprise was that very rarely an alpha wouldn't make it through the foil but would bounce back off it altogether. This, Rutherford famously said later, was "quite the most incredible event in my life . . . as incredible as if you fired a 15-inch shell at a piece of tissue paper and it came back and hit you."

The tissue paper was the sheet of gold foil—an array, therefore, of gold atoms. Those atoms might contain electrons, but an alpha particle could no more bounce backward off an electron than a cannonball can rebound off a Ping-Pong ball. What, then, were the alphas running into?

Most likely, Rutherford already had a pretty good idea what the answer was, but it was a couple of years before he felt confident enough to announce his conclusion. Alpha particles deflected through large angles can only be bouncing off something much heavier than themselves. That something, Rutherford declared in 1911, was the atom's tiny, dense nucleus (a word he introduced the following year).

As with so many great moments in science, this announcement—the birth of nuclear physics—drew little immediate reaction. At an international meeting in 1911 Rutherford said next to nothing, while J.J. described, to no strong interest, a further elaboration of his old plum pudding atom. Rutherford was not a theorist, but he knew that his proposal of the existence of the nucleus left a great deal unsaid. In particular, Rutherford had nothing to say about the atom's complement of electrons. Where were they, in relation to the nucleus, and what might they be doing?

By the time Niels Bohr showed up in Manchester, Rutherford's assistant was Charles Galton Darwin, a grandson of the pioneer of evolution. Darwin was pondering how alpha particles slowed down as they passed through some solid material. It's a rare alpha that hits a nucleus and suffers a great reversal. Mostly, they straggle to a halt, their energy petering out. Darwin's explanation was that they suffer repeated small collisions with the electrons in atoms, losing a little energy each time. By studying this process, he hoped to better understand how electrons arranged themselves in atoms.

He vaguely imagined each atom as having a cloud of electrons that roamed loosely around within a volume representing the atom's overall size. Rutherford's nucleus sat in the middle, somehow keeping the whole thing together. But when Darwin tried to match his model to the measured rates at which alphas slowed in various materials, he came up with atomic dimensions unsatisfactorily different from atomic sizes deduced by more direct means.

In his thesis work, Bohr had conjured up a similarly simple-minded picture of the conduction of electricity by electrons rambling around in metals. That model also did a poor job of explaining what it was supposed to explain. The common flaw, Bohr began to suspect, was that electrons perhaps were not able to move as freely as he and Darwin were assuming.

Somehow, Bohr realized, the nucleus of an atom must keep its complement of electrons in hand, by means of some restraining force. So he imagined each electron not moving freely but held in place, vibrating back and forth, something like a ball on a spring. This was a picture only, a guide to the imagination, but it helped him think.

Now came the momentous but profoundly bizarre step. The electrons, Bohr proposed, could not vibrate with any amount of energy you cared to specify. Instead, they could carry energy only in multiples of some basic "quantum." Now, when alphas passed through some solid material, they could give up their energy to the electrons they encountered only in these quantum amounts.

Remarkably, after a bit of fiddling, Bohr found that he could now give a much better account of how alphas slowed down. Puzzled but satisfied, he wrote up his sketchy theory, sent off a paper for publication, and headed back to Copenhagen to marry Margrethe Nørlund, the sister of friends he had made as an undergraduate.

What remains murky even to this day is why Bohr made this curious suggestion. The idea of a quantum of energy, to be sure, was not new. It had come in 1900, from Max Planck—but in an utterly different context. For years, Planck had been wrestling with an irksome problem. It was well known that hot objects glowed through a series of characteristic colors—from the red glow of embers to the yellow of the sun to the eerie blue-white of molten steel—as their temperature increased. Experimental physicists had carefully measured the spectrum of emitted radiation—a graph of the amount of energy coming out at different wavelengths or frequencies. But theorists had been hopelessly stymied in their attempts to explain the shape of the spectra their experimental colleagues measured.

Almost in desperation, Planck tried dividing the energy of radiation up into little units. It was intended as a mathematical trick, to simplify his calculations. If he could work out the desired form of the spectrum, he supposed, he would then be able to use standard mathematical techniques to shrink his little chunks of energy down to infinitesimal size while keeping his solution intact. His plan half worked. He was able to derive the correct spectrum, but only if he kept the units of energy at a specific size. To his everlasting chagrin, he could not make these quanta, as he called them, go away.

Planck was a conservative sort. There was no reason in standard physics why the energy of electromagnetic waves should be restricted in this way. He refused to believe that electromagnetic energy, in some intrinsic way, could only exist in small units. Rather, he thought, something about the way material bodies emitted energy caused radiation to pop out in discrete quanta. Other physicists mostly agreed with this rationale. Planck struggled mightily in the following years to find a satisfactory reason why energy should emerge in this chopped-up way. He didn't succeed, but neither did he give up.

More than a decade later, Planck's idea remained mysterious and controversial. Still, as Bohr recalled, the idea of energy quanta was at least in the air. It didn't seem extravagantly farfetched to apply a version of the same idea to electrons in atoms. He cheerfully admitted that he couldn't offer any real justification for his proposal. But it seemed to help.

Only some months later did the extraordinary fertility of this innovation begin to dawn. Back in Copenhagen, Bohr had accepted a junior position at the university, the main burden of which was teaching physics to medical students. One day a colleague asked him whether his strange picture of electrons in atoms might be of any help in explaining something called the Balmer series of spectroscopic lines in hydrogen. Sheepishly, Bohr confessed that he didn't know what this Balmer series was, and went to the library to educate himself.

No doubt he knew what spectroscopy was. A century earlier the German astronomer Joseph von Fraunhofer had scrutinized the spectrum of light from the sun and noticed that the rainbow of colors, from red through green to violet, was marked by hundreds of thin dark lines. The spectra of bright stars, he later found, showed similar lines, some coinciding with what he had seen in sunlight, some differing. Over the following decades it was established that each chemical element absorbs and emits light not in a broad, continuous way, but at specific characteristic wavelengths: the harsh yellow of sodium; the companionable red of neon; the ghostly bluish tint of mercury light.

To chemists especially, spectroscopy offered a marvelous diagnostic tool. By looking at the light from some heated sample, they could see what elements it contained. But physicists hadn't come close to understanding why atoms emitted and absorbed only at these characteristic frequencies. It was just another job for the overworked atom to take on.

The Balmer series was the only contribution to science of Johann Balmer, a Swiss schoolteacher. In 1885 he had devised a simple algebraic formula that reproduced with remarkable accuracy the frequencies of a prominent series of spectroscopic lines displayed by hydrogen. But this was pure numerology, devoid of any physical reasoning. In the twenty-seven years that had passed before Bohr became aware of it, no one had come close to explaining where Balmer's formula came from.

But that's exactly what Bohr now did, in the space of just hours. With a mix of physical reasoning and inspired guesswork, he persuaded his sketchy atomic model to cough up the Balmer formula in a few lines of algebra. If Rutherford, a year or two earlier, had given birth to nuclear physics, Niels Bohr had now delivered atomic physics into the world.

Instead of thinking of the electrons as vibrating in some generic way, Bohr now imagined specifically that they orbited the nucleus as the planets orbit the sun. Where gravity holds the solar system together, attraction between negatively charged electrons and a positive nucleus maintains order in the atom. But now Bohr imposed the crucial quantum condition: the orbiting electrons cannot have any energy they like, but can take on only a limited set of values.

If this prescription holds, the single electron of a hydrogen atom must occupy one of a set of distinct orbits. The larger the orbit's diameter, the greater the energy of the electron speeding around. Miraculously, Bohr now saw, his model explained spectroscopy. When an atom absorbs energy, an electron hops from a lower orbit to a higher one; if the electron falls back again, the atom throws back that same dollop of energy. These absorptions and emissions can come only in fixed amounts, dictated by the restricted set of electron orbits. And with appropriate adjustment, Bohr found, these orbits could be placed so as to reproduce precisely the Balmer series. It wasn't merely that he was able to work out a theoretical basis for Balmer's formula. His far greater achievement was that he had at last found the reason why there is a science of spectroscopy at all: it has to do with *transitions* of electrons from one orbit to another.

Tempering his excitement was Bohr's clear understanding that he could give his simple model no persuasive physical foundation. The electrons stayed in their allotted orbits only because Bohr wrote a rule saying that they must. For this restriction, he said plainly in his published paper, he would offer "no attempt at a mechanical foundation (as it seems hopeless)." The model worked beautifully, but where it came from not even Bohr would venture to guess.

To many older scientists, Bohr's atom did not even qualify as physics. Lord Rayleigh, a seventy-year-old mathematical physicist of wide-ranging accomplishment, told his son, "Yes, I have looked at it, but I saw it was no use to me. I do not say discoveries may not be made that sort of way. I think very likely they may be. But it does not suit me." Rayleigh was a thoughtful, modest man, a wise old owl by this time. His opinion of the Bohr atom was not so much a condemnation as a melancholy acceptance that his day had passed.

A shrewd early criticism came from Rutherford, to whom Bohr

had sent a long early draft of his ideas. Rutherford tried to trim the manuscript, to encourage what he called an English habit of terseness in place of continental prolixity, and was surprised by Bohr's stubborn insistence in trying to say everything as fully, carefully, and precisely—overprecisely, Rutherford thought—as he could. Among his comments Rutherford offered this thought: "There appears to me one grave difficulty," he wrote to Bohr. "How does an electron decide with what frequency it is going to vibrate and when it passes from one stationary state to another? It seems to me that you would have to assume that the electron knows beforehand where it is going to stop."

Spontaneity: that awkward notion crops up again. In Bohr's atom, an electron in a high orbit seemed to have a choice of what lower orbit it would jump into, and therefore what spectral line it would produce. In radioactive decay, as Rutherford well knew, a particular unstable atom always disintegrated in the same way, even though the timing of the event was unpredictable. But Bohr's jumping electrons seemed to choose not only the timing of their leap but the destination too. This Rutherford found disturbing.

Nor was Rutherford the only skeptic. Einstein at first cast a wary eye on the new atom. But in 1916 he published a provocative analysis, deceptively simple yet powerfully revealing, which made him think harder about Bohr's achievement. He imagined a single Bohr atom immersed in electromagnetic radiation and asked how the two would exchange energy back and forth. Specifically, he asked how this system would attain thermal equilibrium, with the atom giving out energy as often as taking it in and the spectrum of radiation maintaining a constant form, characteristic of some fixed temperature.

From this simple setup Einstein drew some remarkable conclusions. To begin, the radiation spectrum in equilibrium must have precisely the form Planck had calculated in 1900, with his quantum hypothesis. Next, the atom could give up and take in energy only in units exactly equal to the energy difference between two orbits—meaning that it couldn't, for example, simultaneously shoot out two quanta of lesser energy amounting to the same total.

These conclusions not only bolstered the case that both Planck and Bohr had got their ideas right but also hinted at some deep connection between their proposals. But a third result made him uneasy. For the energy balance between atom and radiation to come out right, Einstein found, the atom's emission of energy had to be governed by a simple law of probability. The chance of the atom shooting out a quantum of energy, he calculated, was constant in any given period of time. He had seen this before. "The statistical law," he noted, "is nothing but the Rutherford law of radioactive decay."

These two processes, in other words—the radioactive decay of a nucleus and the hopping of an electron from one orbit to another—were not only both spontaneous, but spontaneous in the same way. In neither case is there any special time when the change happens—it just happens, for no evident reason. Which appears to mean that these physical phenomena proceed without any identifiable cause.

"That business about causality causes me a great deal of trouble," Einstein wrote to a colleague some years later, when the puzzle still had found no adequate explanation. He was largely alone in his worries. Most physicists were too busy playing with the Bohr atom to spend time fretting over these metaphysical concerns. It would take them a little while to catch up.

Chapter 5

AN AUDACITY UNHEARD OF IN EARLIER TIMES

In July 1914 Bohr took his atom on the road. With his younger brother Harald, an up-and-coming mathematician, he traveled to Germany to present his ideas in Göttingen and Munich. The University of Göttingen, squarely in the middle of the country, was a formidable center of both pure mathematics and mathematical physics. Carl Friedrich Gauss, one of the great mathematicians of all time and a note-worthy physicist, too, had taught there for many years until his death in 1855. But by the early twentieth century, Göttingen had succumbed to the stuffiness that often afflicts great institutions in the wake of a legend (think of Cambridge in the generation or two following Newton). It happened that Harald Bohr had been in Göttingen when his brother's atomic model made its debut, and he reported back to his brother that most of the professors there found the proposal more "bold" and "fantastic" than plausible. One crusty senior mathematician, Harald wrote to Niels, said that "randomly chosen numbers could be made to agree just as well" with hydrogen's spectral lines.

Presenting his theory in person, Bohr made some headway. Not yet fluent in German, he spoke softly and tentatively, but with an unmistakable intensity. The general opinion among the Göttingen faculty, according to a junior physicist by the name of Alfred Landé, was that Bohr's proposal was "all nonsense . . . just a cheap excuse for not knowing what is going on." Max Born, a professor then in his early thirties, had found Bohr's atomic model utterly incomprehensible when he first saw it in print, but after hearing Bohr speak so earnestly in its defense, he told Landé that "this Danish physicist looks so like an original genius that I cannot deny there must be something to it." In just a few years, both Born and Landé would be making their own contributions to this modern theory of atoms.

Bohr had an easier time in Munich, where the head of theoretical physics was the forty-six-year-old Arnold Sommerfeld. Although he had spent a number of years in Göttingen, Sommerfeld retained a youthful zest for innovation and novelty. He had been one of the first to embrace Einstein's special theory of relativity, when other physicists of his generation were struggling to accept that space and time had changed. When the Bohr atom arrived on the scene, he had written promptly to tell Bohr that although he couldn't yet dismiss a certain skepticism about the model, its ability to yield quantitative results was "unquestionably a great achievement." In Munich, Sommerfeld received Bohr warmly, and encouraged his students to turn to the new physics.

It was now August 1914, a fateful month. Niels and Harald Bohr left Germany to hike for a while in the Tyrolean Alps. In the newspapers they read urgent, panicky accounts of impending war and learned that summer vacationers across Europe were streaming homeward across a tense continent. The Bohrs got on a train and found themselves back in Germany just half an hour after the declaration of war with Russia. Reaching Berlin, they encountered screaming crowds, breathless for the fighting to begin. "It is the custom in Germany," Bohr observed drily, "to find such enthusiasms as soon as something military is concerned." After another anxious train journey to the northern coast, they boarded a ferry to Denmark and safety.

Just as Bohr had made his debut in the German physics world, the war closed off most contacts for years. In the meantime, he was trying to find himself a better situation in Copenhagen. He had no laboratory and, burdened with teaching physics to medical students, hardly any time for research. Worse, he had no colleagues with whom he could thrash out his ideas. He started agitating for the university to open an institute of theoretical physics, but with war in the offing the Danish government could put no high priority on such a plan. Instead, Bohr gratefully accepted an offer from Rutherford to return to Manchester. But now Rutherford took up war research (he devised methods to detect submarines by the noise they made underwater), and Bohr was left largely to fend for himself.

Throughout his life Bohr's ideal working method was to involve himself in a continuous, open-ended discussion, a permanently convened informal seminar with colleagues. He thought out loud, threw out ideas, commented and criticized, jumped forward, digressed, stopped and pondered. His two years in Manchester were personally happy for him and his young wife (the industrial city was less charming than Cambridge, she said, but the people were warmer) but scientifically lonely.

Despite the war, science went on. Sequestered in Germany,

Sommerfeld took up Bohr's atom in earnest. Papers and journals trickled back and forth across the trenches. Ideas could still travel. Even indirectly, Bohr could strike sparks in others.

The original Bohr atom explained, really, one thing alone. It accounted for the Balmer series of lines in hydrogen. But there were other lines, other atoms, and even the Balmer lines were not as simple as Bohr at first thought. The American physicist Albert Michelson, using a spectrometer of exceptionally high quality, had discovered in 1892 that individual lines, examined closely, often resolved into doublets—that is, two lines set closely together, corresponding to spectral excitations at two very slightly different frequencies.

It occurred to Bohr that this splitting of spectral lines might arise if electron orbits could be elliptical as well as circular. This happens because the electrons are moving so fast that certain effects of Einstein's relativity become important. In Newtonian mechanics, an infinite family of orbits can exist, all with the same energy but with varying degrees of ellipticity. Each family has one circular orbit, which has zero ellipticity. But relativity makes the energy of all these orbits slightly different, depending on how elliptical they are.

So Bohr imagined that if an atom could contain an elliptical orbit partnered with each circular one, it would then have two slightly different transition energies, depending on which orbit an electron jumped into or out of. And that would cause spectroscopic lines to split into two. But at this point Bohr, alone in Manchester, got stuck. Why would there only be one elliptical orbit, and what would determine its ellipticity? Some new rule was needed, and Bohr couldn't see it.

For a man counted among the great theorists of physics, Bohr had remarkably little ability in the higher realms of mathematics. His papers are not festooned with equations. Instead, he sets out broad concepts and assumptions and tries to draw out quantitative conclusions as simply as possible. Through most of Bohr's career, it was only with the help of a string of mathematically gifted assistants that he was able to turn his remarkable physical insights into quantitative arguments. This way of working fed into the somewhat mystical status Bohr gradually attained. He seemed to be able to discern where the answer to some problem lay, even though he couldn't see exactly how to get there. Many years later Werner Heisenberg wrote of a conversation in which, he said, "Bohr confirmed to me . . . that he had not worked out the complex atomic models by classical mechanics; they had come to him intuitively, rather, on the basis of experience, as pictures."

Unable to work out fully his idea of elliptical orbits, Bohr published a sketchy outline of the suggestion. This paper found its way to Munich, where it came before the highly trained and resourceful mind of Arnold Sommerfeld. Educated in the best German tradition, equipped with a mastery of mathematical techniques and their application to mechanics, electromagnetic theory, and much else, Sommerfeld was just the man to make the next move.

Incorporating Bohr's idea into a sophisticated analysis of the orbital mechanics of the atom, Sommerfeld cooked up a plausible argument to explain why the ellipticity of electron orbits must be restricted to certain values. Ellipticity, like the sizes of the orbits themselves, was "quantized."

Other spectroscopic puzzles yielded to similar reasoning. When atoms are placed in electric or magnetic fields, their spectral lines split into doublets, triplets, and more complicated combinations. These are known as the Stark and Zeeman effects, after their respective discoverers. They came about, Sommerfeld and others now proposed, because electron orbits must lie at some angle relative to these externally imposed fields, and depending on the angle, the orbit energy would change slightly. Here again, not just any old angle was permitted. Orientation too was quantized into a set of allowed dispositions.

In this more complicated system, three so-called quantum numbers were needed to specify any particular electron orbit. The first indicated the orbit's size, the second its ellipticity, the third its orientation. Electron jumps among these various orbits could account for a host of spectroscopic subtleties.

Bohr was thrilled to see the capabilities of his atom expanded so far and so fast. "I do not believe I have ever read anything with more joy than your beautiful work," he wrote to Sommerfeld. So important were Sommerfeld's augmentations that many physicists began to speak of the Bohr-Sommerfeld atom.

These were the triumphal years of what became known as the old quantum theory. It was a funny business, no doubt. The mechanics of the orbits followed entirely from old physics—the electrons obeying Newtonian rules (with occasional modifications from Einstein), controlled by an inverse square law of attraction between electrons and nucleus. But then the quantum limitations came in. Of the infinite range of possible orbits, only certain shapes and sizes and alignments were in fact permitted. These quantum rules had a certain logical consistency, but at bottom they were arbitrary, imposed by fiat.

Conceptually, this awkward hybrid of old and new made little sense. Where did the quantum rules come from? How, as Rutherford had asked, did an electron decide when to jump and where to jump to? Were these jumps in fact triggered in some unknown way, or were they—as Einstein feared—truly spontaneous and ultimately unpredictable?

To these strange, unprecedented questions, no one had the remotest inkling of an answer. But for the time being, no matter!

The Bohr-Sommerfeld atom splendidly explained all manner of hitherto impenetrable spectroscopic mysteries. It did its job inexplicably well, undeservedly well.

The rise of the Bohr-Sommerfeld atom marked not only a maturing of quantum theory but also a historic displacement of the geographical center of theoretical physics from Great Britain to continental Europe, and especially to Germany. The atomic nucleus was a sterling product of the British Empire, conceived by Rutherford, a New Zealander, after work in Canada and England. The primitive Bohr atom could likewise claim a substantial British pedigree, since it derived in large part from Bohr's contact with Rutherford and Darwin. But in the war years, while Bohr stayed in Manchester, his ideas took root in Germany, and that was where the old quantum theory of the atom came to fruition.

Niels Bohr all his life remained devoted to Rutherford, whom he had first met soon after his father died and whom he described as "almost like a second father." Over the years he continued to let Rutherford know how his work on the atom was going, telling him at the beginning of 1918, "I am at present myself most optimistic as regards the future of the theory." Rutherford always responded encouragingly, but at heart he was a practical man, an experimenter. He told his Cambridge colleagues that the quantum theorists "play games with their symbols, but we, in the Cavendish, turn out the real solid facts of Nature." Rutherford liked to say, in his booming manner, that any physicist worth his salt ought to be able to explain his researches to a barmaid, otherwise what was the point? Bohr had trouble enough explaining his physics to his fellow physicists.

But as long as he could still get his ideas across to Rutherford, perhaps he could feel he was on safe ground.

In 1916, with plans for his own institute meeting official approval (and having turned down offers to stay on in Manchester or move to Berkeley, California), Bohr returned to his beloved Copenhagen. There he would found an institute to build quantum theory. But that would take time, and while Bohr wrestled with bureaucracy as well as research, it was Sommerfeld and his students in Munich who took the lead.

In Great Britain, meanwhile, theory went on hiatus. It may have been that the British tradition in mathematical physics, like the empire itself, was overburdened and exhausted. The previous era's giants were gone. The resounding achievements of nineteenth-century Britain, in electromagnetism, optics, acoustics, fluid dynamics, and so on, were a hard act to follow. Some remnant of a Victorian ethos held sway, a spirit of bluff practicality, heartiness, *mens sana in corpore sano*. Theory, in the classical style, ought not to stray too far from common sense. The new ideas of quantum theory—like new art, new music—seemed dangerously avant-garde, unconnected to the plainspoken theories that had worked very well thus far. Experimental physics, especially nuclear physics, flourished in Britain under the powerful command of Rutherford, who in 1919 took over the reins of the Cavendish Laboratory from J. J. Thomson. But theory—deep theory, modern theory—subsided.

Germany, meanwhile, was by no means a blank slate. In both theory and experiment German physicists had built a solid reputation. There had been in the German-speaking world, more-over, a bruising battle over the *meaning* of theory—a debate that most British scientists professed to find amusing, the sort of thing morbidly philosophical Germans might indulge in but not straightforward Anglo-Saxons. Ludwig Boltzmann, a firm believer in the reality of atoms, had clashed with his fellow Austrian the physicist-philosopher Ernst Mach, who was cheerleader in chief for the ideology of positivism. To Mach, theory harbored no deep meaning about the fundamental structure of the physical world. A theory was merely a set of mathematical relationships linking tangible phenomena. The atom, therefore, was at best a convenient fiction, at worst an unverifiable hypothesis.

The atomists had won that battle. Boltzmann's struggles brought him sympathizers and allies among the pure mathematicians, who were intrigued to see physics making incisive use of principles and theorems that had seemed to belong only to them. German theorists, by the early twentieth century, had become mathematically venturesome in a way that their British counterparts were mostly not.

And then there was the First World War, the war to end all wars. At first, it all went very satisfactorily for the Germans, who imagined that German culture and civilization were about to eclipse tired Anglo-Saxon ways. But that expectation imploded in 1918, when the German authorities crumbled and surrendered almost before their people knew that anything was amiss.

In October 1914, when prospects looked glorious, Max Planck had been one of ninety-three distinguished German intellectuals to put their names to an "Appeal to the Cultured Peoples of the World." This lamentable manifesto, published in newspapers across the country, announced the virtue of the German cause, the many superiorities of German civilization, and the tender respect held by Germans for the cultural achievements of lesser nations. What prompted this declaration had been the destruction by German troops of the historic library in Louvain, Belgium. Planck and his fellow intellectuals denied that cultured, civilized Germans could have committed such an outrage, denied reports that Belgian towns and villages had been destroyed, denied, really, that Germany was anything more than an unwilling, put-upon victim of the carnage now spreading across Europe.

Four years later, with the country devastated, the population starving, and flaring socialist revolution provoking reactionary backlash in the anarchic cities, this document became as pathetic as it was shameful. Planck later claimed he had not properly read the appeal when he signed it, but had done so because of the distinguished list of those who had already appended their names. He did, though, during the war, begin to moderate his unthinking embrace of German unity and purpose, and he acknowledged, in response to letters from colleagues elsewhere in Europe, that German soldiers had not always conducted themselves according to the high standards the appeal proclaimed.

Even so, the spirit behind the appeal lived on in a chastened way. Germany might be physically destroyed, but intellectual Germany must endure. The country at war's end was ruined, economically, politically, and psychologically. During the "turnip winter" of 1916–17 people had starved and frozen, and food continued to be short after the war. Political institutions fell apart. Competing factions ranging from extreme monarchists to outright communists indulged in gang violence and assassination. The rest of the world showed no sympathy. Germany had brought about its own ruin. The onerous Treaty of Versailles imposed huge reparations on an already impoverished country. Germany was made into an international pariah, excluded from the budding League of Nations. In the scientific world, Germans were ostracized, refused entry to international conferences, refused publication in many journals.

Amid this dark turmoil, Planck and others believed, science could stand as a beacon for the future. In the *Berliner Tageblatt* at the end of 1919 Planck stated his confidence that "as long as

German science can continue in the old way, it is unthinkable that Germany can be driven from the ranks of civilized nations." Planck was like any number of Germans who, having been at first wholeheartedly in favor of the war, later decided it had been an aberration, a disastrous misadventure imposed by rabid militarists on an unwilling populace. Now that it was all over, Planck thought, German pride and honor and tradition could live on in science. The isolation imposed by the outside world made German scientists all the more determined to save their profession and, with it, some fraction of their country's honor.

That year, 1919, saw the sudden rise to international fame of Germany's greatest theorist, Albert Einstein, whose general theory of relativity received much ballyhooed confirmation in observations, by the British astronomer Arthur Eddington, of the bending of light by the sun's gravity. But Einstein's Germanness was a delicate matter. Born in southwestern Germany and educated for a time in Munich, young Albert had rebelled against the intellectual rigidity and military overtones of his schooling, and at the age of fifteen had fled to Milan, Italy, where his father had gone to establish an electrical business. Later Einstein enrolled at the Swiss Polytechnic in Zurich and moved smartly to obtain Swiss citizenship, renouncing his German passport. By the end of the war, however, his fame had brought him appointment at the center of German science, as a professor in Berlin. Germany, for the time being, proudly claimed him.

In politics as well as science, Einstein was his own man and floated above crass considerations of nationality or chauvinism. He loathed German militarism, but did not approve of the postwar scientific isolation of Germany. It would only prolong hostility and ill feeling, he thought, and he was mostly right. Though he had no love for certain overly patriotic German scientists—Johannes Stark, discoverer of the Stark effect, was soon to take a leading role in denouncing the "Jewish science" of relativity and, later, quantum theory—Einstein stayed away from a number of international meetings on the grounds that all Germans were disbarred, regardless of their politics, attitudes to the war, and current efforts to restore comity.

Einstein's growing worldwide celebrity thrust his political views as well as his authorship of relativity into the public arena. Other of his scientific achievements have tended as a result to be sometimes eclipsed. In the rise of quantum theory, Einstein's crucial role was in turning Planck's mysterious little allotments of energy into physically meaningful units of electromagnetic radiation. In his miraculous year of 1905, two of Einstein's four legendary papers established special relativity (the short second paper included the world's most famous scientific equation, $E = mc^2$). Another, as we know, dealt with Brownian motion. The fourth concerned what he took to calling "light quanta." Einstein argued for taking Planck's argument about little packets of energy at face value: treat the energy packets as if they were bona fide discrete little objects, employ the standard statistical methods developed by Boltzmann and others, and many of the established properties of electromagnetic radiation pop right out. If that failed to convince, he had another argument. By asserting that light was made up of little packets of energy, he was easily able to explain previously puzzling details of the photoelectric effect, in which light striking certain metals generates a small voltage.

But belief in light quanta went against the enormous and continuing success of Maxwell's classical wave theory of the electromagnetic field. What's more, taking light quanta seriously inevitably brought the coupled problems of discontinuity and unpredictability into physics. Classical waves always behaved smoothly, gradually, seamlessly. Light quanta, if such things there were, necessarily came and went abruptly, without apparent reason or cause. Here is the root of a problem that was to plague Einstein for the rest of his life. He believed in the reality of light quanta sooner than anyone else, but he rebelled more strenuously than anyone else against the implication that light quanta inevitably bring spontaneity and probability into physics.

Insisting on the reality of light quanta, Einstein traveled for many years a lonely road. Physicists, meanwhile, puzzled over electromagnetic radiation, radioactivity, the structure of atoms, indeed the structure of basic physics generally. Theorists, Planck dolefully reported in 1910, "now work with an audacity unheard of in earlier times; at present no physical law is considered assured beyond doubt, each and every physical truth is open to dispute. It often looks as if the time of chaos again is drawing near in theoretical physics."

In 1916, in Chicago, Robert A. Millikan carefully measured the photoelectric effect and resoundingly demonstrated that "Einstein's photoelectric equation . . . appears in every case to predict exactly the observed results." Obstinately, though, he concluded that "the semicorpuscular theory by which Einstein arrived at his equation seems at present wholly untenable." Many other physicists, despite the evidence, agreed with Millikan more than with Einstein.

Adding to the confusion, the Bohr-Sommerfeld atom enjoyed only a few years of untrammeled success. It did enough things sufficiently well that it could not be set aside. But as the 1920s dawned, confidence waned that it could do much beyond the simple case of hydrogen, and that only imperfectly. Perhaps, some physicists began to think, this was just a passing phase. Perhaps the disturbing language of transitions and jumps, of quanta and spontaneity, would soon fade away, allowing physics to deal once again in the familiar certainties of old.

At the end of the war, Arnold Sommerfeld took on a couple of interesting new students. In 1918 Wolfgang Pauli arrived from Vienna. Two years later Werner Heisenberg, a local boy, showed up. Unburdened by the past, these young men would quickly make their presence felt.

Chapter 6

LACK OF KNOWLEDGE IS NO GUARANTEE OF SUCCESS

I f Max Planck fervently clung to the culture of science as a way for Germany to rise above the indignity of its downfall, young men like Wolfgang Pauli and Werner Heisenberg found in the pursuit of science a personal escape from the hardships of life in the grim postwar years. Both were children of privilege, sons of university professors. Both enrolled at the University of Munich at a time when that city had survived starvation only to fall into violent anarchy, a cycle of revolution and repression punctuated by assassination. In later memoirs and interviews they do not dwell on these irksome circumstances. For these two young men life meant science, its splendors and frustrations. Science gave them purpose and freedom.

Pauli's origins were especially conducive to his later career.

His father, a professor of medical chemistry in Vienna, was a faculty colleague of Ernst Mach's and something of a disciple of the old positivist. In 1900 he asked Mach to be godfather to his newly born son. The Paulis were by this time Catholic, having converted from Judaism in hopes of securing themselves against the wave of anti-Semitism sweeping across Viennese society. As many as 10 percent of Austrian Jews converted in this period.

Mach was, the younger Pauli said much later, "a stronger personality than was the Catholic priest. The result seems to be that, in this way, I was baptized as 'Antimetaphysical' instead of Roman Catholic." Mach called himself antimetaphysical because he condemned as metaphysics any suggestion that theory could reveal deep secrets of nature, beyond a mere accounting of experimental facts. Pauli could hardly follow his godfather in embracing anti-atomism, but Mach's antimetaphysical severity evolved in him into a kind of universal skepticism, a wariness about theorizing that strayed too far from the concrete and the demonstrable. In the early days of quantum theory, this was a debatable virtue. Heisenberg said later that Pauli wanted to hew strictly to the experimental data *and* maintain mathematical rigor, and in an uncertain and evolving world that was asking too much. Pauli published much less than he might have, Heisenberg said, because so few ideas met his exacting standards. But he was an acute critic and adviser, the "conscience of physics," as he later became known.

At *Gymnasium* in Vienna, Pauli's brilliance in physics and mathematics shone out from the start. Through his father's influence he obtained advanced tutoring from some of the university physics professors, and by the time he graduated he had already written a cogent paper on the new subject of general relativity. When it came to his continuing education, the University of Vienna did not impress young Pauli. Ludwig Boltzmann had committed suicide in 1906, an outcome of his lifelong mix of depression, hypochondria, and self-described neurasthenia, exacerbated by continuing hostility from Mach and the anti-atomists. The Vienna physics department was but a pale imitation of its former self. Pauli had no sentimental affection for the city. Viennese politics was in chaos, society in tatters. The same was largely true in Munich, but the university there at least possessed a thriving and adventurous department of theoretical physics, led by Sommerfeld. In 1918, with the war not yet truly over, Wolfgang Pauli traveled to Munich and signed on as an undergraduate. Diagnosed with a weak heart, he had avoided military service in the last year of the conflict.

Pauli arrived in a country on the point of collapse. In Munich on November 8, the socialist leader Kurt Eisner proclaimed a soviet republic in Bavaria, ousting King Ludwig III. The following day a collection of moderate democrats meeting in Weimar announced the foundation of a new democratic Germany. Two days later came the armistice, when Kaiser Wilhelm in Berlin reluctantly stepped down. No one seemed to be in charge. The right wing wanted to restore the monarchy; the left wing wanted a truly communist Germany. In February 1919 Eisner was assassinated by reactionaries. A second Bavarian people's republic was declared in April, bringing a brief period of red terror as avenging socialists and communists took care of the old regime. Brief, because the militarists returned to crush the socialists two weeks later and embarked on a still fiercer white terror to eradicate the communist scourge.

Heisenberg, then a schoolboy in the city, remembered that "Munich was in a state of utter confusion. On the streets people were shooting at one another, and no one could tell precisely who the contestants were. Political power fluctuated between persons and institutions few of us could have named."

August 1919 saw the promulgation of the Weimar Constitution, a compromised attempt at democracy that pleased hardly anyone. Right-leaning moderates, such as Max Planck, hankered after the civic certainties of the old Germany and regarded democracy as a polite word for mob rule. The left, wanting socialism in earnest, condemned democracy as pitifully anemic. In elections the following year, extremists on both sides did well, while the moderate middle, beloved by no one, fared poorly.

But a fragile, tentative sense of calm slowly returned. Weimar Germany was never truly stable, but Germans gradually gained some confidence that their country would not fall apart the next day. In Munich, the budding scientists Pauli and Heisenberg, having done their best not to notice the chaos around them, found by degrees that they could breathe a little easier.

Sommerfeld, invited to contribute an encyclopedia article on relativity, turned the task over to his precocious new student—"a downright amazing specimen"—who had already written on the subject. In this way Wolfgang Pauli, a mere undergraduate, composed what was in essence a short book on relativity, setting out the mathematics and physics with an elegance and lucidity that astonished Einstein himself.

But general relativity, Pauli soon concluded, was not the subject for him. Though intellectually impressive, it was a finished theory, with no practical consequences. (It would be decades before the language of general relativity became commonplace in astrophysics and cosmology, subjects that didn't exist in 1920.) At Munich, under Sommerfeld's guidance, Pauli could hardly fail to take up quantum theory instead, with its array of cryptic results, unsolved problems, and half-baked theories. He tackled the ionized hydrogen molecule—two nuclei sharing a single electron. This ferociously difficult problem seemed worthy of his attention. He constructed elaborate and ingenious models, trying to figure out how an electron would orbit in this double system, then trying to understand how quantum rules would apply to the orbits. But he made little progress.

He was hooked, though. He began to profess a certain disdain for Sommerfeld's program of sifting through spectroscopic data in order to find patterns that he could interpret as quantum rules. Looking beyond hydrogen and helium to other families of elements in the periodic table, Sommerfeld tried to tease out regularities even in these complex cases. He compiled his findings in a fat monograph, *Atomic Structure and Spectral Lines*—Sommerfeld's bible, as it became known—in which he consciously likened his efforts both to Kepler's search for mathematical and geometrical order in the orbits of the planets and to the old Pythagorean belief in numerical harmonies. "What we are listening to nowadays in the language of spectra," declared Sommerfeld, giving way to a rare flash of purple prose, "is a genuine atomic music of the spheres, a richly proportioned symphony, an order and harmony emerging out of diversity."

Sommerfeld understood that the search for numerical regularities was a way of laying the groundwork for a deeper theory, just as Kepler's laws of planetary motion, derived from close scrutiny of the observed motions of the planets, gained their true meaning only when Newton's inverse square law of gravity gave theoretical foundation to the workings of the solar system. But to the harshly analytical Pauli, Sommerfeld's strategy was an odd combination of theoretical conservatism and latter-day mysticism. Better, Pauli thought, to try building rational theories from sound principles—although his attempt to find such a theory for ionized molecular hydrogen hadn't gotten him very far. The way forward was clear to no one.

Because he developed in Munich a lifelong habit of staying out late at bars and cafés, Pauli generally missed morning lectures. Sommerfeld had firm views about proper conduct and insisted that Pauli get up at a decent hour and work while his brain was fresh. Pauli made an effort to comply, but the habit didn't take, and he reverted to his preferred hours. A tubby young man, Pauli had a tic of rocking back and forth constantly as he sat in his chair and pondered. Sommerfeld concluded that he could not mold his strange, brilliant student into any semblance of what he regarded as normal behavior, and acceded to his late hours and eccentric ways. Pauli referred to Sommerfeld behind his back as a hussar colonel, but to his face showed a lifelong respect and deference that he accorded no one else, not even Einstein.

Sommerfeld was a Prussian by birth, and he looked the part. Short, stocky, and fit, he dressed smartly and had splendid waxed mustaches and a military bearing. Well into his forties he took part eagerly in practices as a reserve army officer. He was a sportsman and an excellent skier. In his youth he had enthusiastically participated in the drinking and dueling that flourished then among student societies.

But Sommerfeld's conservative appearance was deceptive. His mastery of classical physics did not close his mind to innovation. He seized eagerly on Bohr's ill-founded but marvelously productive model of the atom and employed his extensive and detailed knowledge to turn the simple Bohr atom into a sophisticated theoretical device.

Nor in his personality was Sommerfeld the Prussian he seemed to be. With his students he was friendly and collegial. As well as his regular classes he conducted every week an intense two-hour session on the latest research topics. "A kind of market place to exchange views about the most modern developments" was Heisenberg's description of these freewheeling discussions. Sommerfeld's students thus came to learn and criticize at firsthand the ever-changing quantum theory of the atom. He engaged them as contributors to his constantly revised and updated *Atomic Structure and Spectral Lines*. Not just Pauli and Heisenberg but a remarkable number of other contributors to the nascent quantum theory emerged from the Munich school of theoretical physics.

Sometime in 1920 Sommerfeld would have introduced into his weekly research seminar his latest innovation, a fourth quantum number. In the Bohr-Sommerfeld atom up to that point, electrons were described by three quantum numbers that had straightforward geometrical meaning in terms of the size, ellipticity, and orientation of their orbits. But now Sommerfeld took a fateful step away from such commonsense imagery.

The fourth quantum number derived from Sommerfeld's scrutiny of the so-called anomalous Zeeman effect displayed by certain multi-electron atoms. (This is a more complicated version of the original Zeeman effect, the splitting of spectral lines in a magnetic field.) Noticing, as was his habit, certain numerical regularities in the spectroscopic data, Sommerfeld devised a new quantum number that seemed to account for the pattern. But this fourth number had no theoretical foundation; it didn't come with any obvious interpretation in terms of the geometry or mechanics of electron orbits. Straining for a justification, Sommerfeld argued that in these atoms, a single outlying electron took part in all the relevant transitions, while the nucleus and remaining inner electrons formed a composite, invariable core. The whole thing thus looked like a modified kind of hydrogen, and Sommerfeld suggested that the fourth quantum number involved what he vaguely called a "hidden rotation" of the single outlying electron relative to the core.

To Pauli, this was not theory but fantasy. It was one thing to take standard properties of electron orbits and transform them into quantum numbers. It was quite another to invent a quantum number out of whole cloth and only afterward grace it with some dubious, ad hoc interpretation. Did Sommerfeld's new invention imply that the quantum atom had properties that couldn't be understood by reference to old-style mechanics? Or did it just mean that quantum theory was coming off the rails?

It might have been around this time that Pauli suggested caustically to Heisenberg that "it's much easier to find one's way if one isn't too familiar with the magnificent unity of classical physics. You have a decided advantage there," he told his fellow student with a wicked grin, "but then lack of knowledge is no guarantee of success."

If Pauli arrived in Munich almost as a mature, fully formed physicist, armed with not only deep knowledge but also pronounced opinions, Heisenberg was by contrast talented but dreamy, with a spotty command of his subject. He had thought at first to take up pure mathematics, but in his teens he discovered a small book written by Einstein as an attempt to explain relativity to nonscientists. "My original wish to study mathematics," he recalled later, "was imperceptibly diverted toward theoretical physics."

Werner Heisenberg was born at the end of 1901 in the university town of Würzburg, some 150 miles northwest of Munich, where his father taught classics. August Heisenberg was devoted to Bismarckian Germany, a Protestant nation united in moral conduct and the pursuit of commerce. His family lived with proper decorum. They went to church dutifully and regularly, though August later confessed to his two sons that he had never had any particular religious sensibility. Late in life Werner said, with an elegant ambiguity befitting the inventor of the uncertainty principle, that "if someone were to say that I had not been a Christian, he would be wrong. But if someone were to say that I had been a Christian, he would be saying too much."

In 1910, August Heisenberg was appointed professor of Byzantine philology at the University of Munich, and the family moved to the Bavarian capital. Professor Heisenberg was a good teacher but a fierce disciplinarian. Trapped within his rigid, formal manner lay a volatile temper that flashed out occasionally, usually within the privacy of his family. He pushed Werner and his older brother, Erwin, to compete with each other, in athletics and scholastics, and Erwin mostly had the edge. Only in mathematics, Werner discovered, could he beat Erwin, and this discovery became the foundation for his life. Werner and Erwin were never close. After studying chemistry, Erwin moved to Berlin and fell in with the cult of anthroposophy. As adults, the brothers had only rare and fleeting contacts.

Finishing *Gymnasium* just as the war was ending, Werner had to serve in the local militia, a ragtag collection of teenagers charged with keeping order in the strife-torn city. It was like playing cops and robbers, he said later; nothing serious. He remembered times "when our families had long since eaten their last piece of bread," when he and his older brother and other friends would scurry about the shattered city of Munich foraging for food. During the time of the Bavarian soviet, he had sneaked across the battle lines into territory controlled by forces of the German republic, returning with bread, butter, and bacon. Such memories Heisenberg recounted in a matter-of-fact way, as if these adventures had been the stuff of a perfectly ordinary adolescence.

He was a shy, careful child. During the war, his character began to emerge. Charged with adult responsibilities in the local militia, Werner discovered a certain charisma, an ability to lead and command respect, if not affection. Away from his straitlaced family, he found room to breathe among loose organizations of young men who tramped the mountains, hiked the countryside, and lost themselves in earnest adolescent discussions of art and science, music and philosophy. Such groups went back a couple of decades and belonged to larger associations with such names as *Pfadfinder* (pathfinder) and *Wandervogel* (migratory bird). Modeled on the Boy Scout movement recently started by Baden-Powell in Great Britain, the German groups tended to be more romantic in spirit than their hearty, practical British counter-parts. After the war especially they became a repository for all kinds of wistful, wishful thinking about a new and peaceful soci-ety. As Heisenberg put it, "the cocoon in which home and school protect the young in more peaceful periods had burst open in the confusion of the times, and . . . by way of substitute, we had discovered a new sense of freedom."

The youth movement was at heart adolescent and middle-class, an indulgence available only to the fortunate. Thomas Mann, in *Doctor Faustus*, described similarly earnest bucolic pil-grimages by young students and commented sharply that "such a temporary style of life, when a city-dweller who engages in intel-lectual pursuits becomes a stopover guest at some primitive rural spot of Mother Earth's . . . has something artificial, patronizing, dilettantish about it, a trace of the comic."

In some of these youth organizations lay seeds that would grow into the strident and violent Hitler Youth of a decade or so later. But Heisenberg's group remained assiduously apolitical, and his wanderings (there were trips as far afield as Austria and Finland) represented for him a solace that he clung to even as his scientific career blossomed. All his life Heisenberg wanted to believe that he could deal with the nasty exigencies of political strife by looking the other way and retreating into nature.

In 1920 Heisenberg's father arranged for Werner to interview with Ferdinand Lindemann, an elderly mathematics professor at Munich. Years before, Lindemann had opposed Sommerfeld's appointment on the grounds that an applied mathematician who dabbled in physics was a wretched creature indeed. He occupied a gloomy office crammed with furniture of dated design. On the desk sat a small black dog, which glared at the young supplicant and proceeded to yap ever more loudly as Lindemann attempted to probe Heisenberg's interests and knowledge. Through the din, Heisenberg managed to confess nervously that he had been reading about relativity. "In that case you are completely lost to mathematics," said Lindemann, concluding the interview.

So Heisenberg went to see Sommerfeld instead and got a warmer, though not uncritical, reception. Sommerfeld was impressed by Heisenberg's command of mathematics and his interest in current physics, but disturbed by the candidate's apparent concern for philosophical questions rather than the scientific fundamentals of experiment and theory, which Heisenberg seemed to find insufficiently grand. Walk before you run was the essence of Sommerfeld's advice: if you want to tackle the deep questions, you have to gain mastery of the subject first. Heisenberg went away thinking that physics might turn out to be a little tedious. With his yough movement friends he wrangled over big issues: What is knowledge? How can we be sure of it? What constitutes progress? Sommerfeld wanted him to learn about the fine structure of spectral lines in hydrogen and the anomalous Zeeman effect in the alkali metals. Nonetheless, Heisenberg signed on to study physics with Sommerfeld.

For his thesis work, he took up a safe problem in the classical physics of fluid flow, but that was a mere sideshow compared to his rapid immersion in quantum theory. Heisenberg was not nearly as well schooled in physics as Pauli, but perhaps for that very reason he was less hidebound, less apt instantly to see the difficulties rather than the possibilities of strange but promising suggestions.

Pauli told Heisenberg that once you had the right mathematics, you had all you needed; you could pose problems and calculate answers. But Heisenberg wanted something more, a more elemental or visceral understanding. Of the quantum atom they were trying to elucidate, he told Pauli, "I have grasped the theory with my brain, but not yet with my heart." The Bohr-Sommerfeld atom of those days was, he said, a "peculiar mixture of incomprehensible mumbo-jumbo and empirical success."

But this mumbo-jumbo was clearly the most exciting part of physics to be involved in. Sommerfeld introduced Heisenberg to the fourth quantum number he had recently devised, and asked his new student to see if he could expand the scheme to embrace more of the oddities of the anomalous Zeeman effect. Ingenious and resourceful, showing both technical skill and scientific imagination, Heisenberg did what his teacher asked—and came up with a result that startled them both. In trying to account for a greater variety of spectral lines, Heisenberg devised a clever formula that worked nicely as long as he gave the already mysterious fourth quantum number half values: $1/2$, $3/2$, $5/2$, and so on. (It would do no good to multiply by 2 to get rid of the fractions, because then the sequence would be 1, 3, 5 . . . missing the even numbers.)

This Sommerfeld was not prepared to contemplate. A half quantum went against the whole point of the enterprise. Pauli agreed. Once you allowed halves, you would open the door to quarters and eighths, he said, and soon there would be no quantum theory left.

While Heisenberg and Sommerfeld wrestled over this bizarre proposal, they were taken aback to see essentially the same idea put into print by another young German, Alfred Landé. Landé had first learned of quantum theory as a student in Göttingen, during Niels Bohr's prewar visit. Like Heisenberg, Landé offered no justification for the half-quantum trick, except that it seemed to account for a couple of interesting puzzles.

Irked by his loss of priority, Heisenberg now tried to recapture the lead by attempting a theory of the half quantum. Sommerfeld had suggested that the fourth number had something to do with the rotation of an outer electron relative to the atom's core. Heisenberg insouciantly went further and proposed that this rotation could somehow be split into half units, one part belonging to the electron, the other to the core. When this outer electron made a transition, only a half quantum of rotation would come into play.

Heisenberg was enraptured by his own ingenuity, but neither Sommerfeld nor Pauli took to his idea. It was adventurous and imaginative, to be sure—or speculative and unfounded, to put it another way. Even so, Sommerfeld consented to let this paper go to a journal, where it became Heisenberg's first published work. Landé didn't think much of the idea either, and wrote to Heisenberg pointing out that his theory effectively threw away the sacred principle of conservation of angular momentum. Heisenberg didn't much care. All the old rules were up for grabs. As Landé put it many years later, Heisenberg's strategy, when he ran up against a difficult problem, was not to look strenuously for a solution within the confines of known physics but immediately to search for something wholly new, something radical. This attitude would bring the young man great success, but it could misfire too.

Sommerfeld likewise judged Heisenberg truly smart but

alarmingly cavalier. In a word, immature. He thought enough of his young pupil's work to write to Einstein about it, praising Heisenberg's attempted theory but admitting to reservations. "It works fine, but the foundation of it is quite unclear," he said. "I can only press on with the technicalities of quanta; you must make your philosophy."

Remarkable or foolish or both, Heisenberg's first attempt at theoretical physics upended his earlier attitude. He now saw that progress came not from pondering weighty philosophical issues but from trying to solve specific problems. And it was good to keep one's mind open to new thinking. Pauli's jibe had some truth. Heisenberg didn't know enough physics to see how absurd his half-quantum theory was. But then, Heisenberg already suspected, Sommerfeld was inclined to be too cautious, and Pauli too skeptical. Many years later Heisenberg met the great American physicist Richard Feynman, who lamented that young physicists were no longer allowed the luxury of making mistakes. Their teachers and colleagues jumped ferociously on any unsound reasoning before it had a chance to blossom. But sometimes, Feynman told Heisenberg, he might have an idea that he knew didn't make sense, but "damn it, I can see that it's right."

Under Sommerfeld's guidance, Heisenberg had the profoundly valuable experience of seeing his first idea in physics, inspired but controversial, thrown into the open and left to fend for itself. It was exhilarating. Criticism only spurred Heisenberg to keep at it. He had found his path. The classical order was disintegrating, and Heisenberg would join the search for a new system. In physics as in politics, the young man had no nostalgia for old certainties.

Chapter 7

HOW CAN ONE BE HAPPY?

During the summer of 1922, Germany enjoyed a momentary calm. Food was scarce, but few starved. Money was tight, but the hyperinflation that obliged people to cart around billions of marks' worth of scruffy notes in wheelbarrows to buy bread and milk had not yet caught fire. In Göttingen, the weather was gorgeous, and it was there that theorists gathered in June to listen to a series of lectures on quantum theory from the subject's acknowledged guide and master, Niels Bohr. Sommerfeld naturally went, and he insisted that his precocious and already controversial pupil Heisenberg come too. Even in the relatively well-to-do Heisenberg family there was little spare money to go around, so Sommerfeld paid for Werner's trip to Göttingen out of his own pocket. Heisenberg slept on someone's couch and was constantly hungry. But that was nothing unusual for students in those days, he recalled.

Pauli was also there. After taking his doctorate at Munich the previous autumn, he had spent the winter semester in Göttingen, then had moved on to a position in Hamburg. Now he journeyed south to meet Bohr for the first time.

Bohr's visit was significant politically as well as scientifically. Like Einstein, Bohr detested German militarism and imperialism but did not approve of the postwar attempt to isolate German science from the rest of the world. A furious insistence on vengeance would not engender peace.

Bohr had already begun to reestablish contacts in Germany. He had visited Berlin in 1920 at the invitation of Planck and Einstein. It was Bohr's first meeting with these two towering figures, and both found the young Dane admirable. Einstein and Bohr subsequently exchanged little mash notes. "Not often in life has a person delighted me so by his mere presence," wrote Einstein to Bohr. "I am now studying your great papers and—if I happen to get stuck somewhere—I have the pleasure of seeing your friendly young face before me, smiling and explaining." "It was one of the greatest experiences in my life to meet you and speak with you," Bohr responded. "I shall never forget our conversation on the way from Dahlem to your home."

Two years later, when Bohr visited Göttingen, some of the old stuffiness of that university had dissipated. The new head of theoretical physics was Max Born, who eight years earlier, during Bohr's prewar visit, had been one of the eager young scientists in the back of the audience. Born possessed a measure of the Göttingen fondness for mathematical rigor, but he had embraced the wondrous new physics despite its sloppiness and inconsistency.

In the lovely June weather of 1922, Bohr delivered, in his discursive, Delphic style, a series of lectures setting out the view of quantum theory from Copenhagen. The Bohr Festspiele this luminous week was later dubbed, echoing a Handel Festspiele that played in Göttingen about the same time.

The windows were open, and summer noises drifted into the hushed seminar room. One local member of the audience complained that the senior Göttingen faculty members, as before, bagged the best seats at the front, leaving the junior scientists to gather at the back, where they struggled to catch Bohr's slow, indistinct words. But Heisenberg was entranced. He had learned his quantum physics from Sommerfeld, whose technical style emphasized simple models and elementary calculations. Of the master's voice, by contrast, Heisenberg said that "each one of his sentences revealed a long chain of underlying thoughts, of philosophical reflections, hinted at but never fully expressed . . . It all sounded quite different from Bohr's lips."

Bohr talked about some recent ideas that he and an assistant had been developing in Copenhagen. Heisenberg, who had read the papers and criticized them with Pauli, had the temerity to speak up from the back of the room with his objections. This caused the gentlemen at the front to turn their heads. Bohr knew of Heisenberg by name, from his work on the rather detestable half-quantum idea, and after the lecture was over he invited the young man on a long walk. They strolled up the Hainberg, the modest hill that overlooks Göttingen, and, sitting in a coffee shop, dissected quantum theory. "My real scientific career only began that afternoon," Heisenberg said years later.

He wanted to know, Heisenberg told Bohr, what quantum theory meant. Beyond the ingenious calculations and fitting of complex spectral lines to peculiar systems of quantum numbers and rules, what, he wanted to know, was the underlying conception, the true physics of it all? Bohr did not insist on the need for detailed classical models that could be translated systematically into quantum terms. Rather, he told Heisenberg, the point of models was to capture as much as one could hope to say about atoms, given the inadequacies of the ideas with which physicists were fumbling along. "When it comes to atoms," Bohr concluded enigmatically, "language can be used only as in poetry. The poet, too, is not nearly so concerned with describing facts as with creating images and establishing mental connections."

This, to Heisenberg, was strange and revelatory. Only a generation earlier, Boltzmann and his allies had argued strenuously for the atom as a concrete *thing*, not a theoretical abstraction, still less a poetical allusion. Was Bohr now saying that physicists couldn't hope to describe atoms concretely, that they must make do with analogies and metaphors? That the intrinsic reality of an atom was inaccessible to them? That perhaps it was meaningless even to talk about the intrinsic reality of an atom?

It is not clear how far the reader can trust Heisenberg's account of these and other meetings. Writing many years after the events, he pretends to reconstruct long, intense conversations, set out in complex, thoughtful paragraphs. It is hard to shake the feeling that Bohr, in Heisenberg's recollection, says things that suit a view of Bohr's physics fashioned by Heisenberg over the many intervening years. What's undeniable is that Heisenberg's first meeting with Bohr truly changed his view of what quantum theory was about.

Although Bohr understood that quantum theory might not follow classical rules, he also insisted, from the earliest days, that the language of classical physics—which so successfully described the everyday world—remained indispensable. His bridge across that gap was an overarching idea he called the correspondence principle, which said that the quantum theory of the atom ought to match seamlessly on to classical analyses of atomic behavior, when the latter are known to work. For example, electron jumps between low-lying orbits, close to a nucleus, involve large and abrupt changes of energy, whereas in transitions between states with large quantum numbers—distant orbits in the far reaches of an atomic system—the energy change is small compared with the energies of the orbits themselves. The more modest the quantum jump, the more closely it resembles the kind of incremental change amenable to classical treatment. The correspondence principle meant that in such instances, quantum and classical behavior should tend to the same outcome, and indeed Bohr had used reasoning of this sort to flesh out details of his atomic model.

In general, though, putting the correspondence principle to good use in complicated situations demanded a certain finesse on the part of the practitioner. A textbook published in the early 1920s said that the correspondence principle "cannot be expressed in exact quantitative laws [but] in Bohr's hands it has been extraordinarily fruitful." Abraham Pais, who has written extensively on this period of physics, comments mysteriously that "it takes artistry to make practical use of the correspondence principle." Emilio Segrè, another physicist reminiscing about the old days, agrees that the correspondence principle was hard to formulate precisely, and explains that in practice it amounted to saying, "Bohr would have proceeded in this way."

Thus arises the Bohr mystique. In his schematic, intuitive way, Bohr saw how to build quantum theory, and other physicists were supposed to follow his lead, even though they couldn't quite see what he was doing. Bohr acquired a reputation for lecturing in slow, rambling, painfully constructed sentences pregnant, so it seemed, with great meaning that lay just a little beyond the audience's reach. Somehow it was always the listener's responsibility to discern what Bohr meant, not Bohr's job to speak more clearly. Like any guru worthy of the name, Bohr was cryptic and indirect.

Writing home to his parents after his first encounter with Bohr, Heisenberg evidently felt he had made an impression. Although Bohr as well as Sommerfeld expressed grave reservations about the half-quantum idea, they had to concede, according to Heisenberg, that they couldn't prove him wrong and that their objections came down to "generalities and matters of taste." Bohr apparently described Heisenberg's work in one of his lectures as "very interesting," which the young man, as yet unfamiliar with Bohr's idioms, took as an endorsement. At the end of it all, Bohr indicated to Heisenberg that he should find a way to spend some time in Copenhagen.

Bohr had a new disciple.

In a sign of the rapidly increasing maturity of the American scholarly scene, Sommerfeld was invited to spend the academic year beginning in September 1922 in far-flung Madison, Wisconsin. He was happy to spread the quantum gospel to an eager, fresh audience, and the chance to earn some foreign income as the German mark became ever more worthless was nothing to scoff at. While he was away, he arranged for Heisenberg, who had not yet graduated, to continue his studies with Born in Göttingen.

In the interim, Heisenberg went to the annual meeting of the Society of German Scientists and Physicians in Leipzig in September, where he hoped in particular to meet Einstein. But anti-Semitism and the campaign against Jewish science were gaining momentum. In June, shortly after Bohr's triumphant lectures in Göttingen, right-wing militants in Berlin gunned down the German foreign minister, Walther Rathenau, a Jew and a friend of Einstein's. Workers, trade unionists, and socialists organized and protested. Rightist groups in turn shouted louder against communists and Jews. In this delicate, dangerous atmosphere, Einstein chose not to go to Leipzig.

Heisenberg's visit was eye-opening. At the first session he attended, a leaflet was thrust into his hand, which proved to be a circular from the German science movement decrying the polluting influence of Jewish thought. In his memoirs Heisenberg professed to be shocked at this intrusion of coarse politics and prejudice into the tight world of science. But he could hardly have been unaware of these vicious hatreds. His shock was that he could no longer wish them away, or pretend they were some sort of transient aberration that would collapse under the pressure of reason. Scientists could be as irrational and vituperative, as opportunistic and selfish, as the mobs in the streets. Science was not the citadel Heisenberg dreamed of.

After that first session, he returned to his lodgings to find that everything he had brought with him had been stolen. Left only with the clothes he wore and the return half of his train ticket, he skipped straight back to Munich and thence, a little later, to Göttingen. There, at least, he could hope to find sanctuary in a university town that took pride in its intellectual detachment from the travails of the world outside.

Pauli had spent the previous winter semester at Göttingen. Born wrote to Einstein that "young Pauli is very stimulating—I shall never get another assistant as good," but he was miffed to find he had to send a maid to get Pauli out of bed at 10:30 every morning. Nor did Pauli's brusque independence and sharp tongue endear him to the quiet and formal Born. Pauli made snide reference to the excessive brand of rigor and pedantry he called *Göttingen Gelehrsamkeit*—Göttingen scholarliness. Years later Born said of Pauli, "I was, from the beginning, quite crushed by him . . . He would never do what I told him to do—he did it his own way, and generally he was right."

Though Born, as the aging Sommerfeld began to retire from the front lines, would oversee an equally influential school of quantum theory in Göttingen, he never attained the general respect and affection that Sommerfeld inspired. He had been a shy, sensitive child, easily discouraged by small slights, and grew into a reserved, rather timorous, occasionally peevish adult. His original intention to be a pure mathematician had faltered when, after a brief undergraduate spell in Göttingen, he felt overawed by the mathematical talent around him. Moving into physics, he proved adept and versatile—a dilettante, to use a word he applied to himself—but always remained both diffident about his abilities and quick to take offense if his contributions went unnoticed. During the war years he was appointed a professor in Berlin, where he became close to Einstein as the general theory of relativity burst upon the world. "I was so impressed by the greatness of his conception," Born wrote later, "that I decided never to work in this field."

He became a good teacher and mentor but, as his experience with Pauli illustrates, he could be cowed by students sharper and more confident than he felt himself to be. Unlike Pauli, Heisenberg proved capable of getting up in the mornings unaided, and showed proper respect. He was, Born recalled, "quite different; he was like a little peasant boy when he came, very quiet and friendly and shy . . . Very soon I discovered he was just as good in brains as the other one."

From Born, Heisenberg learned yet a third attitude toward the development of quantum theory. Sommerfeld forged ahead by solving problems, troubled little by either mathematical nicety or philosophical profundity. Bohr tried to force vague concepts and dimly perceived suggestions into rational shape, and only then started looking for a mathematical formulation. Born, by contrast, was reluctant to say anything that he couldn't yet express in a formal mathematical way. Though he had abandoned any desire to be a true mathematician, his thinking retained a powerful strain of the mathematician's desire for strict reasoning and watertight logic.

In Göttingen, traces of the old ethos remained. Observing the increasing use of rarefied mathematics in physical theory, David Hilbert, the presiding mathematical genius, made the not-quite-joking remark that physics was becoming too difficult for physicists—the implication being that only mathematicians could be trusted to do the job properly. Born at least half agreed. He didn't share Bohr's belief in the importance of working out the concepts first. "I always thought mathematics was cleverer than we are—one has first to find the correct formalism before one should philosophize about it," he said. Heisenberg formed a distinctly different view. "Born was very conservative in some ways," he said. "He would only state things which he could prove mathematically . . . [He] had not so much feeling about how things worked in atomic physics."

That was Born's unfortunate role: to the physicists, too much the mathematician; to the mathematicians, not enough.

Still, Heisenberg acquired a new degree of mathematical sophistication from his time with Born, who conducted a regular seminar at his house with half a dozen eager students. But even in these early days, as a mere undergraduate judging an established professor, he was far from convinced that Born had the right kind of imagination to push science forward.

Under Born's guidance, Heisenberg tried to apply his ideas, including the half-quantum system, to neutral helium—two electrons orbiting a doubly charged nucleus. Spectroscopically, helium displayed all kinds of complications. It had both single and multiple lines, and when electric or magnetic fields were applied, those lines split in hopelessly complex ways. Heisenberg and Born concluded before very long that they could not understand helium at all, even with all the augmentations and ornamentations of the Bohr-Sommerfeld atom that were now floating about. The same conclusion emerged from Bohr's institute.

Meanwhile Alfred Landé, having already beaten Heisenberg to the punch with the half-quantum business, rolled out a further elaboration in which he added more peculiar rules to produce a scheme that mimicked yet more curiosities of the Zeeman effect. Pauli despaired of this strategy. He couldn't deny that Landé's tricks and devices appeared to fit various complicated sets of spectroscopic data, but as far as the search for an underlying theory was concerned, these efforts struck him as fatuous.

Having taken a position in Hamburg, Pauli quickly excused himself to spend some months in Copenhagen, where he could learn quantum theory from Bohr. One day, Pauli recalled, he was stumping about the streets when a friend came across him and said he looked glum. "How can one be happy when one is thinking about the anomalous Zeeman effect?" Pauli responded smartly, and went on his way.

For all his earlier enthusiasm over Sommerfeld's elaborate atomic models, Bohr was increasingly unimpressed by the Munich game of mindlessly fitting quantum numbers and odd numerical systems to all manner of spectroscopic lines. Such efforts brought no real enlightenment, but seemed rather to degenerate into mere tinkering, in which each new spectroscopic puzzle was answered by some arbitrary theoretical adjustment. To Heisenberg and Pauli too it frequently seemed that a line had been crossed; a model can stand only so much ornamentation before its conceptual integrity falls apart. Heisenberg recalled that "some of us had begun to feel that the earlier successes of the theory might have been due to the use of particularly simple systems, and that the theory would break down in a slightly more complicated one."

It was as if physicists, attempting to uncloak the capricious nature of the quantum atom, were resorting to irrationality themselves.

Chapter 8

I WOULD RATHER BE
A COBBLER

In September 1923, Niels Bohr made his first visit to North America, speaking at Harvard, Princeton, Columbia, and elsewhere and concluding with a series of six lectures at Yale. This event *The New York Times* found noteworthy enough to report, though it didn't manage to spell the speaker's name right. "Dr. Nils Bohr," the story ran, would explain "his theory of the structure of the atom, which has been accepted by many scientists as the most plausible hypothesis yet put forward." A helpful subhead added: "He pictures the atom with nucleus corresponding to sun, and electrons to planets."

By this time, of course, the idea of the atom as a miniature solar system was barely tenable even as a loose analogy. At Yale, Bohr described the history of theories of the atom, explained how spectroscopy had become the essential tool for probing the modern atom's structure, talked of how electrons were supposed to inhabit and move within atoms, and hinted at the numerous puzzles theorists currently faced. In words reported by the *Times*, Bohr confessed his inability to clearly describe the quantum atom in familiar language: "I hope I have succeeded in giving an impression that we are dealing with some sort of reality—a kind of connecting up of experimental evidence, with the prediction of new experimental evidence. Of course we cannot offer a picture of the same kind as we have been used to in natural philosophy. We are in a new field where we find that the old methods do not help, and we are trying to develop new methods."

Despite occasional attention from the newspapers, Bohr would never acquire the celebrity and glamorous aura that came Einstein's way. The previous year Bohr had won the Nobel physics prize for his insight into the structure of atoms, but even then he had been overshadowed by Einstein, who was at the same time awarded the delayed 1921 prize. There had been no shortage over the years of Nobel nominations for Einstein, but the Nobel committee, a cautious outfit, was slow to embrace relativity, which still had vehement critics and for which direct evidence remained meager. Einstein almost won the prize in 1920, but last-minute doubts and reservations led the committee to reward instead Charles Guillaume of Switzerland, who had invented a nickel steel with a low coefficient of thermal expansion, cited for its great utility in precision measuring instruments. Einstein's prize, when it finally came, was for his theory of the photoelectric effect, which Millikan's experiments had verified a few years earlier, even though Millikan himself refused to accept that his results demonstrated the reality of light quanta.

The Nobel awards to Bohr and Einstein highlighted a glaring contradiction. Einstein, as he had done for many years now, accepted at face value the reality of light quanta, but then was unhappy at the way they contaminated physics with elements of discontinuity and chance. In sharp contrast, Bohr had invented an atomic model that explained how atoms emitted and absorbed dollops of light at specific frequencies, but then ran into trouble because he refused to accept that these packets of light were truly fundamental to physics.

Just a few weeks later, there came news of an experiment that seemed to settle the question. At Washington University in St. Louis, Arthur Compton succeeded in bouncing X-rays off electrons and found precisely what the quantum model predicted. When a quantum of radiation hits an electron, it bounces off with less energy. But Planck's rule says that the energy per quantum is proportional to the radiation's frequency, so reduced energy means lower frequency or longer wavelength. Compton's careful measurements bore this prediction out. "This remarkable agreement between our formulas and the experiments can leave but little doubt," he concluded, "that the scattering of X rays is a quantum phenomenon."

Sommerfeld, teaching in Madison at the time, relayed the news to Bohr, and as he traveled about America giving lectures on quantum theory, he urged upon his listeners the importance of the experiment. Compton's decisive findings appeared in May 1923 in the American *Physical Review*, now the world's preeminent journal of physics but one that Europeans in those days barely knew. (Heisenberg, interviewed in 1962, recalled that in the early days no one in Germany read the *Physical Review* because of course it didn't exist back then; in fact it was already three decades old.)

Compton scattering stands in the history books as the crucial evidence that light quanta had to be taken seriously. Probably the majority of physicists, like Sommerfeld, reacted to the announcement with enthusiasm and gratitude. Others were more grudging in their acceptance. Niels Bohr's reaction, though, went beyond skepticism into outright hostility. With a stubbornness bordering on blockheadedness, he insisted more strenuously than ever that light quanta could not possibly be real and spent a year working up a sketchy theory of atomic emission and absorption that denied them any role. This episode reveals the dark side of Bohr's character. Convinced that he alone could see the truth, he was intransigent, overbearing, and immune to reason.

Bohr's antipathy to Compton's discovery was, it later emerged, not purely a matter of scientific judgment. He reacted fiercely for the simple reason that he had heard and dismissed the same idea several months earlier, when his own assistant in Copenhagen figured out the theory of what became known as the Compton effect. Bohr had angrily squelched the idea then, and so was instantly ready to do battle when Compton made his announcement.

Bohr's assistant was Hendrik Kramers, a native of Rotterdam. In 1916, Kramers had shown up on Bohr's doorstep in Copenhagen, equipped with a degree in physics and an eagerness to learn quantum theory. The match proved perfect. A quick study and a sharp mathematician, Kramers had a capacity to grasp Bohr's opaquely articulated thoughts and turn them into quantitative theoretical statements. And he could lecture clearly. Only a couple of years after arriving in Copenhagen, Kramers became an informal emissary for Bohr, speaking persuasively to audiences that were often still reserved and skeptical. Kramers offered precise arguments and specific calculations, not the obscure philosophical musing that Bohr favored.

"Bohr is Allah and Kramers is his prophet," pronounced Wolfgang Pauli, notwithstanding that he liked Bohr's assistant a good deal. Kramers, proud and a little insecure, could be prickly and sarcastic. Pauli detected a congenial spirit.

Bohr encouraged Kramers to look into a question that had thus far received little attention. If the most notable characteristic of spectral lines was their wavelength or frequency, a second obvious quality was their intensity. Some lines are brighter than others. The germ of an explanation could be found in Einstein's prescient 1916 paper, in which he showed that atomic transitions followed a law of probability identical to Rutherford's probability rule for radioactive decay. The more probable a transition, Bohr suggested to Kramers, the brighter the corresponding spectral line ought to be.

But Einstein's analysis of the abrupt, probabilistic way in which atoms emitted light also provided further reason to believe that light quanta were genuine physical entities. Following in these footsteps, Kramers could not help but absorb the same lesson.

Sometime in 1921, according to a story unearthed only recently by his biographer Max Dresden, Kramers must have been thinking about the way a light quantum would interact with a particle such as an electron. In short order, he came up with the pleasingly simple collision law that Compton would soon employ to such great effect. In the recollection of his wife, a singer who had acquired the nickname Storm on account of her tempestuous personality, Kramers came home one day "insanely excited." The next day he took his momentous discovery to Bohr. And then, Storm recalled, Bohr went to work on her husband, explaining and insisting and maintaining over and over, in any number of different ways, that the idea of a light quantum was untenable, that it had no place in physics, that it would mean throwing away the hugely successful classical theory of electromagnetism, that it simply would not do. Bohr wouldn't let up. Against Kramers's straightforward calculation, Bohr could put up any number of weighty but elusive arguments, physical, philosophical, historical in nature. Bohr had the trick of being powerfully persuasive even when he was not entirely reasonable. Whenever, by dint of his impressive but inscrutable reasoning, he saw the right answer ahead of all the mathematicians and calculators, his reputation as the mystic of quantum theory only grew. When he was equally relentless in pursuit of a mistaken idea, he could be a bully, plain and simple.

So great was the pressure that Kramers became ill, taking refuge in the hospital for a few days. By the time he came out, he had yielded utterly to Bohr's will. Kramers suppressed his discovery of what would soon be called the Compton effect, to the extent of destroying his notes. He became as vehement as Bohr, if not more so, in his denunciation and ridicule of the light quantum. When Compton published his results, Kramers further repressed the knowledge that he had already calculated exactly what Compton had now disclosed to the world, and joined with his boss in looking for a way to continue the fight against an unacceptable conclusion.

Bohr's adamancy on this point remains genuinely mysterious. It seems to have become fixed in his mind that accepting the existence of discrete light quanta would fatally undermine the wave theory of classical electromagnetism. Others, notably Einstein, saw well enough that there was a basic mismatch between the two points of view but decided this was a problem physics would have to set aside for the time being, until all these new ideas were better assimilated.

Bohr and Kramers, at any rate, set themselves to salvaging their viewpoint. A third young collaborator was drawn into this web. John C. Slater, after earning a doctorate from Harvard, set out in the fall of 1923 on a European tour, stopping in Cambridge for a few months before moving on to Copenhagen. Like most younger physicists, Slater embraced light quanta without reservation, but while in Cambridge, the ancestral home of the classical theory of radiation, he saw dimly how it might be possible to live with light quanta while not throwing away all the undeniable successes of light waves. Both must exist, he thought. He imagined a radiation field, roughly along classical lines, but repurposed. It existed to guide light quanta around and to facilitate their dealings with atoms.

Arriving in Copenhagen, Slater found his embryonic hypothesis warmly received. Bohr and Kramers particularly seized on the suggestion of an underlying field that somehow interacted with atoms, determining how and when they would emit or absorb light. They were not so keen, however, on Slater's idea that the radiation field would also guide the passage of light quanta. They went to work on the young visitor, reiterating by ceaseless tag-team argument how his clever idea could be reshaped into an acceptable theory. The three of them began working on a paper together. That is, Bohr mused out loud, Kramers jotted notes as best he could, and Slater stood expectantly by. In letters home, Slater said how thrilled he was that his ideas were being taken seriously by no less than Bohr. He was confident he would see the finished paper fairly soon, he added. By the end of January 1924 it had been sent out for publication—an astonishingly fast piece of work for anything with Bohr's name on it. Bohr, Kramers, and Slater was the order of the authors.

Characteristically, the BKS paper offers not a tightly constructed quantitative model but a nonmathematical sketch, an outline of a possible theory. It contains but one exceedingly simple equation. Instead, the paper describes in purely qualitative terms a new kind of radiation field that surrounds atoms, influences their absorption and emission of light, and also transports energy between them.

There's another new ingredient, not original to BKS but adapted from an earlier suggestion. As Bohr had explained to his audience at Yale, the idea of electrons orbiting nuclei in planetary style could no longer be taken seriously, but no one had come up with any better account. So BKS used a subterfuge. They pictured an atom as a set of "virtual oscillators," each one corresponding to a particular spectroscopic line. In elementary terms, all simple oscillators—a pendulum, a weight on a spring, an electron zipping around and around—obey basically the same mathematical law. To avoid specifics, BKS made use of the standard physics of oscillating systems without trying to connect the presumed oscillators to some explicit picture of how electrons actually moved within an atom. This was all in keeping with the spirit of BKS, which clearly aimed to offer a blueprint of a possible theory, not a finished model.

A summary sentence from the BKS paper expresses both the vague nature of their proposal and the frustratingly elusive style of Bohr's prose: "We will assume that a given atom in a certain stationary state will communicate continually with other atoms through a time-spatial mechanism which is virtually equivalent with the field of radiation which on the classical theory would originate from the virtual harmonic oscillators corresponding with the various possible transitions to other stationary states."

Bohr seems to think, as lawyers do, that punctuation can only create ambiguity. What's also remarkable, on close inspection, is how nebulous this language is. Crucial arguments are expressed in conditional tenses and rely on locutions of deliberately vague intent: *communicate with, time-spatial mechanism, virtually equivalent . . .* The finicky care with which each phrase was written and rewritten and written again is apparent, yet the curious result is that the more carefully Bohr tries to express himself, the more his meaning recedes. As Einstein once put it, Bohr "utters his opinions like one perpetually groping and never like one who believes he is in possession of the definite truth."

This was meant as praise, evidently, but a later collaborator admitted that Bohr's manner had its downside: "You could never pin Bohr down to any statement; he would always give the impression of being evasive, and to an outsider who didn't know him, he would make a very poor showing."

Among the maddeningly vague proposals BKS freely dispensed, one blunt conclusion stood out: according to their theory, energy was not absolutely conserved. Because the emission and absorption of energy run according to rules of probability, energy can disappear from one place and reappear somewhere else—or vice versa—without the one event being strictly connected, by old-fashioned cause and effect, to the other. The mysterious radiation field acts as a sort of escrow account for energy, so that the sums always add up in the long run, but in the short term there can be temporary deposits and overdrafts.

So eager was Bohr to banish all reference to Einstein's light quanta, because he wanted to preserve classical wave theory, that he ended up throwing classical energy conservation out the window instead. Clearly, there would be no easy reconciliation of these contradictory ideas.

In an odd display of timorousness, probably because he knew what the answer would be, Bohr didn't approach Einstein directly but asked Pauli to find out what the old man thought of BKS. "Quite artificial" and even "*dégoûtant*" (he used the French word) was how Einstein judged the proposal, Pauli reported, adding for good measure that he also completely disapproved of it. And to Max Born, Einstein wrote, "I would rather be a cobbler or even a casino worker than a physicist" if this was where theory was headed. Born himself, asked many years later about BKS, turned the question back to his interviewer: "Can you explain to me what the BKS theory was? It was a thing I never grasped properly in my whole life."

It had but a brief existence. Bohr, Kramers, and Slater were obliged to argue that Compton's results proved only a statistical truth. Individual collisions between X-rays and electrons would not necessarily conserve energy, but in bulk any discrepancies would cancel out. But new experiments, by Compton and others, quickly proved this assertion false. Individual collisions obey precisely the expected rule and conserve energy exactly.

By the spring of 1925, Bohr admitted that BKS was a bust. Slater remained embittered all his life at the way his idea had been mangled, he said later, into something he didn't truly endorse. For Kramers, the failure of BKS, following on the forcible suppression of his discovery of Compton scattering, seemed to mark the end of any ambition that he would one day produce truly great physics. He fell into a mild depression, according to his biographer, and was thereafter more subdued in the employment of his scientific imagination.

The BKS proposal marks, despite all this, a turning point. Depending on one's interpretation of what the theory actually was, it was either the last gasp of attempts to rest quantum theory on some sort of classical foundation or else the first proof that all such efforts were doomed.

The most influential ingredient of BKS, in retrospect, was one that entered into the argument—not unlike Planck's original proposal of the energy quantum—as a sort of trick to get around other difficulties. That was the use of ill-defined virtual oscillators as a means to talk about how an atom emitted and absorbed light while deliberately avoiding any discussion of what precisely the electrons in the atom were doing.

Developing this idea, Kramers proved a little later—in a rigorous mathematical way, as contrasted to the woolly conceptualizing of BKS—that the oscillator picture was far more than a convenient dodge. An atom's interaction with light of any frequency, Kramers demonstrated, could be calculated in its entirety from the appropriate set of virtual oscillators. All the necessary physics was in there.

But did that mean that the old imagery of electron orbits could be dispensed with altogether? Kramers, apparently, thought not. The virtual oscillators were merely an interim substitute, he believed, for the details of an underlying atomic model that would work on more or less traditional lines.

Others took an opposite view. Writing to Bohr, Pauli posed the crucial issue: "It seems to me the most important question is this: to what extent it's allowable to speak at all of definite orbits of electrons . . . In my view Heisenberg has taken exactly the right position on this point, in that he doubts it's possible to speak of definite orbits. Kramers has thus far never admitted to me any such doubt as reasonable."

Seeing Kramers's theory of virtual oscillators, Heisenberg had indeed quickly perceived its revolutionary implications, and just as quickly determined to release the idea from its traditional moorings. It was he who would transform this bold conceptual innovation into a wholly new theory of atoms—in fact, of physics.

Chapter 9

SOMETHING HAS HAPPENED

When Sommerfeld came back from Madison in the spring of 1923, Heisenberg returned to Munich from Göttingen to finish his doctorate. To that end he had pursued a project in mathematical fluid dynamics, unrelated to quantum theory but a steady topic. His doctoral examination was nonetheless a struggle. Because he had to show mastery of physics in general, experimental as well as theoretical, Heisenberg had grudgingly enrolled in a laboratory course under the supervision of Wilhelm Wien, professor of experimental physics at Munich. Wien was a distinguished researcher whose careful measurements of the spectrum of electromagnetic radiation had been crucial to Planck's 1900 introduction of the quantum hypothesis. But the curmudgeonly Wien, conservative in science as well as politics, was skeptical about Planck's innovation and openly detested the quantum theory of the atom that his colleague Sommerfeld was forging.

Wien was thus naturally disposed to show some hostility toward Sommerfeld's latest wunderkind, and the young man's ill-concealed disdain for experimental matters only made things worse. At Heisenberg's oral exam in July, Wien pelted the candidate with questions about his laboratory work that he should have been able to answer easily enough but, through his own neglect and indifference, was not. Wien wanted to know the resolving power of a certain optical device. Heisenberg couldn't recall the textbook formula, tried to work it out on the spot, and got it wrong. Wien was appalled. Only after tense negotiation with Sommerfeld would he reluctantly affirm that Heisenberg had shown an adequate knowledge of the broad range of physics. The brilliant young man got his doctorate, but with a grade barely above a mere pass.

Momentarily abashed, Heisenberg went quickly to Göttingen to confirm that his previously agreed plan to spend the following year there was still acceptable to Born. It was, and Heisenberg immediately left for Finland with his Pfadfinder comrades to refresh his spirit among the northern lakes and forests. By September he was in Göttingen, eager to put his doctorate behind him, along with all pretense of any interest in experimental physics, and apply himself to the perplexing array of puzzles that threatened to bring quantum theory to a standstill.

Heisenberg was gathering clues. In March of the following year, he made a short visit to Copenhagen—his first—where he found Bohr and Kramers in the thick of their enthusiasm over the BKS proposal. Though he balked at that idea in its entirety, one part of the sketchy theory—the virtual oscillators—lodged in his brain. There was at this time not even a halfway reasonable account of how electrons in atoms behaved. So it seemed like an ingenious stratagem, and perhaps more than just a stratagem, to put aside all those irksome technical concerns about electron orbits, and instead think of the atom as a collection of oscillators tuned to the appropriate spectroscopic frequencies.

No one could say what these oscillators were supposed to be, in detailed physical terms. But that was precisely the point. The oscillators were intended to capture the observed characteristics of atoms, not their internal structure, which—as Bohr had been cryptically hinting for some time—might not be amenable to model building in the traditional way. By thinking in terms of these oscillators, theorists gave themselves some breathing room.

Back in Göttingen, meanwhile, Born was hatching his own plan. He published a paper calling for a new system of "quantum mechanics"—the first appearance of that term—by which he meant a structure of quantum rules obeying their own logic, not necessarily following the time-honored dictates of classical, Newtonian mechanics. For Born, sketchy hypotheses and far-flung analogies of the BKS type were no good. He relied on mathematics to light the way ahead, and he had a particular trick in mind.

The language of classical physics is the differential calculus devised by Newton and independently by Leibniz to deal with continuous variation and incremental change. But in trying to understand the workings of atoms, physicists came up against phenomena that were abrupt, spontaneous, and discontinuous. An atom was in one state, then it was in another. There was no smooth passage between the two. Traditional calculus could not cope with such discontinuities. So Born, making a virtue of necessity, proposed instead to substitute a calculus of differences—a mathematical system that would take for its basic elements the differences between states rather than the states themselves.

This, Heisenberg could see, bore some relation to what Kramers was doing with his virtual oscillators. Both approaches brought the transitions between states to center stage and pushed the underlying states into the wings. Digesting these ideas, Heisenberg came up with an ingenious argument that justified theoretically one of the peculiar half-quantum formulas he and Landé had divined empirically some time ago. A small step, of uncertain significance, but perhaps a step in the right direction.

Then came a lull. During the summer of 1924 Heisenberg went off again with the Pfadfinder, this time to Bavaria. Bohr, worn out by the years of effort establishing and now running his new institute, took the summer off to relax in the Swiss Alps and at his country cottage outside Copenhagen. Rutherford, no lazy-bones, commended Bohr for having the sense to take a long break. For the remainder of 1924, and into the following year, quantum mechanics stayed hidden.

Einstein had already told Born he would rather be a cobbler than deal with the kind of physics Bohr, Kramers, and Slater were peddling. He was not the only one threatening to quit the business. Writing to a colleague in May 1925, Pauli moaned that "right now physics is very confused once again—at any rate it's much too difficult for me and I wish I were a movie comedian or some such and had never heard of physics. I only hope now that Bohr will save us with some new idea." (Charlie Chaplin's movies were all the rage in Germany at the time.)

But Bohr had his own ideas about where the urgently needed new idea might come from. "Now everything is in Heisenberg's hands—to find a way out of the difficulties," he remarked to an American scientist visiting Copenhagen about this time.

In September 1924, Heisenberg at last went to Copenhagen for an extended stay of several months. He chose to arrive at a time when he knew Kramers would not be there. Older by seven years, and in manner and appearance older still, Kramers was the only young physicist who intimidated Heisenberg a little. Heisenberg played the piano; Kramers played cello *and* piano. Heisenberg struggled to master Danish and English; Kramers spoke several languages with ease. He was not merely knowledgeable but opinionated too. Where Pauli thought Kramers sharply amusing, Heisenberg found him patronizing. Kramers, Heisenberg said years later (one can almost hear the gritted teeth), "was always a perfect gentleman in every way; he was too much of a gentleman."

And, of course, Kramers had been with Bohr for years, a close relationship that Heisenberg could only envy.

In Copenhagen, Heisenberg started on a research project but soon tangled with both Bohr and Kramers, who scrutinized every publication emerging from the institute and presented Heisenberg with a long list of the deficiencies of the paper he wanted to send out. Heisenberg was "completely shocked," he recalled. "I got quite furious." But he fought back. In personal matters he may still have been shy, but in defense of his science he was truculent and determined. He beat back the objections and wrote a paper that Bohr agreed (not without a further exasperating round of revision) should be published. Heisenberg gained confidence from this experience. He also learned that he might be wise sometimes to keep his ideas to himself for a while.

A paragraph from Pauli to Bohr, written just before Heisenberg's first trip to Copenhagen, is worth quoting for its insight into Heisenberg's scientific character:

> Things always go very oddly with him. When I think about his ideas, they strike me as dreadful and I curse to myself over them. He is quite unphilosophical—he pays no attention to working out clear principles or connecting

them with existing theory. But when I speak to him I like him very much and I see that he has all kinds of new arguments—in his heart, at least. So then I realize—apart from the fact that he's personally such a nice fellow—that he's really outstanding, a genius even, and I think he can truly move science forward again . . . Hopefully you and he together can take a big step forward in atomic theory . . . Hopefully also Heisenberg will come home with a philosophical attitude in his thinking.

So he did, at least a little. During a brief, awkward collaboration with Kramers, Heisenberg took to heart ever more strongly the picture of an atom as a set of tuned oscillators. His perspective on what a theory of atoms ought to do was evolving rapidly. Forever gone now was the old Sommerfeld style of model, with electrons following well-defined orbits governed by classical mechanics. Of course, Heisenberg had nothing yet to put in place of such thinking. But his focus was inexorably shifting. Worry less about what atoms *are*. Think more about what they *do*.

But a shifting perspective is only helpful if it leads to a real theory. Heisenberg had to find some way to give logical shape to his evolving thoughts. Back in Göttingen again, turning the matter over and over, he found a way forward by jumping into the past. What his roving mind now latched onto was the century-old mathematical machinery of the Fourier series.

In the classic but pertinent example, any vibration of a violin string, no matter how harsh or discordant, is equal to some weighted combination of the string's pure tones, its fundamental and harmonics. Heisenberg was already thinking of an atom as a set of oscillators. Now it occurred to him to take that imagery to its fullest conclusion. "The idea suggested itself," he said in a lecture three decades later, "that one should write down the mechanical laws not as equations for the positions and velocities of the electrons but as equations for the frequencies and amplitudes of their Fourier expansion."

Heisenberg's bland phrase doesn't begin to convey the bizarre and radical nature of what he was aiming to do. In classical physics, a particle's position and velocity are its defining characteristics, the basic elements to which the laws of mechanics applied. For electrons in atoms, however, Heisenberg was now proposing to make the frequencies and intensities of the still-hypothetical oscillators the primary elements of a new calculus, so that the position and velocity of electrons would then be defined only secondarily, in terms of the oscillator strengths. This was a revolutionary reversal. Starting with Bohr and Sommerfeld, the central idea of old quantum theory had been to figure out how electrons move in an atom and deduce from those motions the atom's spectroscopic frequencies. Heisenberg turned this logic exactly backward. The characteristic frequencies would be the basic elements of his atomic physics, and the motion of electrons would be expressed only indirectly.

"The idea suggested itself" was how Heisenberg put it years later—but it suggested itself to him and to no one else. Heisenberg's leap here is reminiscent of the leap Einstein made when, by reexamining the apparently self-evident notions of time and location, he was led to his theory of relativity. A judicious questioning of the obvious may well be a mark of genius.

But genius also requires fortitude. It was not difficult for Heisenberg to write down, in a formal mathematical way, equations that expressed an electron's position and velocity as combinations of an atom's fundamental oscillations. But when he inserted these composite expressions into standard equations of mechanics, what he created was an almighty mess. Single numbers became lists of numbers; straightforward algebra exploded into pages of confused, repetitive formulas. For weeks Heisenberg tried different calculations, played algebraic games with Fourier series, floundered uselessly, then ground to a halt when a monstrous attack of hay fever clogged his brain.

On June 7 he took an overnight train to the northern coast of Germany. So red and inflamed was his face that the landlady of an inn where he stopped for breakfast the next morning thought he had been beaten up—not an outlandish possibility in the Germany of the mid-1920s. He then boarded a ferry for the small, barren island of Helgoland, about fifty miles out into the North Sea. A military outpost during the First World War, Helgoland was by this time a resort, frequented by those in search of fresh sea air and isolation.

Heisenberg stayed for a week and a half, clambering about the rocky shore, resting, reading Goethe, talking to hardly a soul but thinking, always thinking. Refuge for Heisenberg always meant a retreat to nature, to mountains, forests, and water. Slowly his head cleared. In this lonely place he could let his mind dwell on physics.

What had brought Heisenberg to a dead stop was not any grand conceptual puzzle but a basic problem of multiplication. He had turned position and velocity from single numbers into multicomponent sums. Multiplying two numbers together produces another number. Multiplying two lists of numbers together creates a page full of possible terms, consisting of each member of the first list multiplied by each member of the second. Which terms were important, and how should they be added to generate a meaningful product?

Wrangling this mess into order, Heisenberg found his answer by concentrating on physics, not mathematics. The elements of his algebra were oscillations, each representing a transition from one state to another. The product of two such elements, he saw, must represent a double transition, one state to a second, then from the second to a third. The way to arrange his multiplication table, Heisenberg now deduced, was to put together elements corresponding to the same initial and final state, summing over all possible intermediaries. This realization—after a little work, to be sure—gave him the key by which he could devise a multiplication rule that was both manageable and sensible.

At three o'clock one morning, lying sleeplessly on his bed in a small hostel, Heisenberg knew that he had the tool enabling him to perform calculations in his new mechanics. He could write down, for example, a mathematical formula for the mechanical energy of some system, expressed in his strange calculus. There was no guarantee he would get a useful answer. His elaborate method might deliver gibberish.

So he rose from his bed and started figuring. In his feverish state he made endless slips and errors and had to start over again and again. But finally he got an answer, and it was more than he could have dreamed for. With joy and bewilderment he discovered that his strange mathematics indeed yielded a consistent result for the energy of a system—but only so long as that energy was one of a restricted set of values. His new form of mechanics was, in fact, a quantized form of mechanics.

This was remarkable but altogether inexplicable. In all previous attempts at the quantum theory of atoms, the physicist had to plug in, somewhere along the way, Planck's original quantization rule or some close variant of it. Heisenberg had done no such thing. He wrote down the standard equations for a simple mechanical system, inserted his strange composite expressions for position and velocity, applied his novel rule of multiplication— and found that the transformed mathematics held together only when the energy took on certain values.

His system, in other words, quantized itself, with no further prompting from him. As Planck, a quarter of a century earlier, had seen that radiation must be quantized, so now Heisenberg, in an utterly different way, had discovered that the energy of a mechanical system must likewise be quantized. This was as wonderful as it was mystifying.

Elated, unable to sleep, Heisenberg went out to the shore in what was now the early light of morning and climbed onto a rock while the sun rose on a new day. What he had found was a gift from above, he thought, a discovery of unwarranted and unexpected proportions. He lay on the rocks in the warming sunlight, marveling at the beautiful consistency of his strange calculations, and thought to himself, he recalled later, "Well, something has happened."

One thing disturbed him. His multiplication rule was not reversible. That is, x times y was not necessarily the same as y times x. This was nothing Heisenberg had ever encountered before. But it was what he needed; it was what the new physics demanded.

Passing through Hamburg on the way back to Göttingen, Heisenberg consulted excitedly with Pauli, who urged him to write up his ideas quickly. In letters to Pauli in the following weeks, Heisenberg complained that things were going slowly, that it was all very unclear to him, that he didn't know how it was going to turn out—but at the same time he passed on to Pauli his latest results, a set of ideas and conclusions that would form the backbone of his developing view of quantum mechanics. By early July, he had written what he called a "crazy paper" setting out his discovery. He sent a copy to Pauli, eager for his friend's judgment but wary too. He was convinced, he told Pauli, that in doing away with the classical notions of position and velocity, he was on the right track; he was still not sure that his transformed versions of these things were right. That part of the paper, he confided, seemed "formal and feeble; but perhaps people who know more can make something reasonable out of it." He begged Pauli to respond to his draft in a couple of days, because "I must either finish it or burn it."

In Göttingen, Heisenberg presented a draft to Born, saying he didn't trust his own judgment enough to know whether it was worth submitting for publication. Born was immediately enthusiastic and sent the paper off to the *Zeitschrift für Physik*. To Born's mathematically acute mind, Heisenberg's strange calculus, awkwardly expressed, provoked surprise, excitement, and an elusive glimmer of recognition that he couldn't at first trace. He conveyed the news to Einstein a few days later, warning him that although Heisenberg's work "looks very mystical, it is certainly correct and profound."

Chastened by his experience in Copenhagen, Heisenberg waited until the end of August before letting Bohr in on the news. "As Kramers has perhaps told you, I have committed the crime of writing a paper on quantum mechanics," he wrote, uninformatively. Kramers, by chance, had been in Göttingen for a few days when Heisenberg returned from Helgoland. He and Heisenberg talked, evidently, but Kramers relayed nothing of their conversation to Bohr. It's entirely possible that Heisenberg, still unsure of his ideas and already wary of Kramers, said too little for Kramers to grasp.

Heisenberg began his paper with a bold declaration. "An attempt is made," he wrote, "to obtain foundations for quantum-theoretical mechanics based exclusively on relationships between quantities that are in principle observable." Observability: it was the coming principle of this new mechanics. Forget about trying to account for the behavior of electrons directly; instead, express what you would like to know in terms of what you can see—the spectroscopic characteristics of an atom.

For all its revolutionary implications, Heisenberg's paper was a curiously abstract presentation. It talked only of simple mechanical systems defined in formal terms. Nowhere was there discussion of actual atoms and electrons. It was a foundation for quantum mechanics, not the thing itself. Whether this new approach would lead to a genuine physical theory remained to be seen.

Pauli, writing to another physicist some weeks later, said that Heisenberg's idea "has given me new *joie de vivre* and hope . . . it's possible to move forward again." Einstein, when he saw the short paper, had a very different reaction. He wrote immediately to a colleague that "Heisenberg has laid a large quantum egg. In Göttingen they believe it (I don't)."

Perhaps Heisenberg had indeed, as he put it, committed a crime in writing on quantum mechanics. The verdict was not yet in. In any case, as he soon discovered, he was not the only perpetrator.

Chapter 10

THE SOUL OF THE OLD SYSTEM

I n November 1924, the science faculty of the University of Paris gathered to hear a doctoral thesis defense. The candidate, Louis de Broglie, was thirty-two years old, having been delayed in his scientific career first by family tradition and then by the war. The de Broglies, over the generations, had provided France with a succession of statesmen, politicians, and military officers. Louis's father was a member of parliament, and Louis had studied history at the Sorbonne with a view to becoming a diplomat. But he had a considerably older brother, Maurice, who got caught up in the 1890s mania for X-rays and decided, against the wishes of their father and grandfather, on the life of a scientist. Maurice filled his younger brother's head with compelling talk of radiation and electrons. Louis too switched to science.

During the war, the younger de Broglie served with a mobile radiotelegraphy unit, learning firsthand the practical value of classical electromagnetic wave theory. From his brother he heard about the controversial notion of light quanta. He was hardly the only scientist to be aware of the seeming inconsistency of these two views of light, but he came at the problem from an angle no one else had thought of.

Late in 1923 an elementary idea crossed his mind. If light, in the form of Einstein's quanta, could act in ways that made it look at least notionally like a stream of particles, might not particles also display some of the properties of waves?

Cobbling together a makeshift but ingenious argument that combined Planck's quantization rule for radiation with Einstein's famous $E = mc^2$ for moving objects, de Broglie was able to set out a logically consistent case associating a wavelength with any speeding particle. The faster the particle, the smaller this wavelength.

But was this more than a mere algebraic formula? Did the implied wavelength actually connote any physical wavelike behavior? Unencumbered by any deep understanding of quantum theory, de Broglie applied his idea to the hopelessly outdated Bohr atom and hit upon a striking result. For an electron circling the nucleus in the innermost orbit, he calculated a wavelength exactly equal to the orbit's circumference. For an electron in the next orbit—higher in energy, with a bigger radius—he found that the circumference was twice the electron's wavelength. The third orbit was three wavelengths around, and so on, in simple progression.

Just as the fundamental note and harmonics of a violin string correspond to those vibrations for which a whole number of wavelengths fit into the string's length, so the allowed orbits of the Bohr atom were those for which a whole number of electron wavelengths fit around the orbit's circumference. Perhaps quantization was no more mysterious, after all, than the physics of vibrating strings.

De Broglie published his idea in two papers that appeared toward the end of 1923. They attracted little attention. A year later, presenting a more complete version at his thesis defense, he got a wary response. His examiners found the notion of electron waves too simplistic and at the same time too fantastic. They couldn't argue with his algebra. Whether it meant anything physically they couldn't decide. Still, one of the examiners sent a copy of de Broglie's work to Einstein, who was fond of simple ideas with huge implications. His verdict was unambiguous. The fog has begun to lift, he commented.

But no one else took much notice.

Born in 1892, de Broglie was a decade older than Heisenberg, Pauli, and the other youthful adventurers who were creating *Knabenphysik*—lads' physics—at Göttingen and elsewhere. Older still was Erwin Schrödinger, born in Vienna in 1887 to an affluent, somewhat raffish family with English as well as Austrian ancestry. An only child, Erwin grew up in a splendid apartment in central Vienna. The Schrödingers had little taste for music but a passion for the racy, erotic theater of late-nineteenth-century Vienna. Erwin was raised by women—his delicate mother and her two sisters. Even at *Gymnasium* he stood out as much for his confident, charming, slightly louche manner as for his evident intellectual capabilities.

Schrödinger enrolled at the University of Vienna in the autumn of 1906, in the weeks following Boltzmann's suicide. Later, during the war, he saw fighting and won a medal. His most influential teacher, Fritz Hasenöhrl, died in battle. Hasenöhrl's demise was one of the chief reasons Pauli, a few years later, left his hometown to study in Munich. After enjoying a number of infatuations during his twenties, Schrödinger married in 1920 a woman who adored him and looked after him. He came to depend on her, as a substitute for the women who had raised him, but saw no reason why marriage should impede his instincts. He eventually had three children with three different women, but none with his wife.

In 1921 Schrödinger accepted a comfortable position in Zurich, where life was much easier than in postwar Vienna. By this time he had published work on electron theory, on the atomic properties of solids, on cosmic rays, on diffusion and Brownian motion, on general relativity—all of it well regarded, none of it spectacular. Although he worked on contemporary problems, Schrödinger was something of a traditionalist. He found repellent the idea that electrons, in the Bohr-Sommerfeld atom, jumped abruptly from one orbit to another. This sort of discontinuity, he thought, did not belong in physics, because—as Einstein also complained—it brought with it a degree of unpredictability, of things happening for no discernible reason.

As word began to leak out of de Broglie's reinterpretation of electron orbits as standing waves, Schrödinger realized that a theoretical result of his, published a year or two earlier, had in a far more opaque way been hinting at the same thing. Where de Broglie was a dabbler theoretically, Schrödinger was equipped with a sophisticated mastery of mathematical techniques. He latched onto de Broglie's intuitive sketch with the idea that it ought to be possible to make a real theory out of it.

In the middle of 1925, when Heisenberg was on his rocky outpost in the North Sea devising his peculiar new calculus, Schrödinger wrote a paper enlarging on de Broglie's electron

waves, in which he threw out in passing the suggestion that particles are not really particles at all but, as he put it, "whitecaps" of
an underlying wave field. That suited Schrödinger's view of the
physical world. Once you accept the existence of particles, of discrete packets of energy, you cannot avoid discontinuity, spontaneity, and all such related ills. But if what we think of as
particles are in reality the superficial manifestation of underlying
waves and fields, then continuity may be restored.

As Schrödinger incessantly talked up the wonders of de
Broglie's waves, a more skeptical Zurich colleague challenged
him. If these were, so to speak, the waves of the future, then
where was the wave equation? De Broglie's argument merely af-
fixed a wavelength to an electron moving with a certain speed. It
said nothing about what these waves were, what determined
their form, what if anything was their physical meaning. For
all respectable classical wave motions—electromagnetic waves,
ocean waves, sound waves—a mathematical equation relates the
thing oscillating to the force or influence that makes it oscillate.
For de Broglie's waves no such equation existed. They were not,
at this point, actual waves so much as a disembodied or ab-
stracted idea of wave motion.

Over the Christmas holiday of 1925, Schrödinger took himself away from his wife and spent some days at a resort near
Davos, Switzerland, with a girlfriend whose name has been lost
to history. In what one physicist later described as "a late erotic
outburst in his life," Schrödinger (close to forty years old by now)
found what he was looking for—a wave equation that captured
de Broglie's intuition in a formal manner. (In truth, this was but
one of many erotic outbursts in Schrödinger's life, although it is
the only one that gave rise to great physics.)

Schrödinger's equation described a field governed by a mathematical operator that embodied a kind of energy function. Ap-

plied to an atom, the equation yielded a limited number of solutions in the form of static field patterns, each one representing a state of the atom with some fixed energy. Quantization came about in what seemed like a pleasingly classical way. To obtain representations of atomic states, Schrödinger stipulated that the solution ought to go to zero, as the mathematicians say, at large distances—otherwise it wouldn't correspond to an object localized in space. With this condition, his equation yielded up only a finite set of stable configurations, each possessing some discrete amount of energy. This was no more mysterious, he thought, than getting a finite set of vibrations for a violin string fixed in place at both ends.

Even better, Schrödinger hinted in one of the several papers he published in 1926, it might now be possible to understand a quantum jump, a transition from one state to another, not as an abrupt and discontinuous change but as a fluid transformation of one standing wave pattern into another, with the wave reconfiguring itself rapidly but nonetheless smoothly.

The old guard was delighted. Einstein wrote enthusiastically to Schrödinger, scribbling in the margin of his letter that "the concept of your paper shows real genius." He and Planck quickly invited Schrödinger to Berlin. Einstein wrote again to tell Schrödinger, "I am convinced that you have made a decisive advance . . . just as I am convinced that the Heisenberg-Born way is going in the wrong direction." Classical order, it suddenly seemed, had been restored.

What Einstein called the Heisenberg-Born way was by contrast an exotic, intricate, forbidding mathematical system that had blossomed quickly from Heisenberg's Helgoland inspiration. On

July 19, Max Born, still struggling to resolve the glimmer of familiarity that Heisenberg's strange calculus had evoked, had taken a train to Hannover for a meeting of the German Physical Society. As he sat in his compartment reading and scribbling, recognition dawned: what Heisenberg was doing, in his extemporized way, belonged to an arcane branch of mathematics that went by the name of matrix algebra. Born remembered learning something about it years ago, when he still had thoughts of becoming a pure mathematician. Until now he had never seen it put to any practical use.

A matrix is an array of numbers set out in rows and columns. Matrix algebra is a set of arithmetical rules for combining and manipulating matrices in a systematic fashion. The elements of Heisenberg's calculations could likewise be written out, Born now saw, in the form of square arrays, with each position in the array denoting a transition from one atomic state to some other state. Crucially, the multiplication rule that Heisenberg had so painfully devised was precisely the multiplication rule for matrices already known to a select band of mathematicians. Heisenberg had known none of this, of course. It was his acute insight into physics that led him to the answer he needed.

Born now realized that an entire branch of mathematics already existed, ready-made for quantum mechanics. At some point Pauli, coming down from Hamburg, joined the same train and came across Born thrilled with his discovery and eager to explain what he now understood. Pauli was not just unimpressed but floridly caustic. "I know you are fond of tedious and complicated formalism," Born recalled him saying. "You are only going to spoil Heisenberg's physical ideas by your futile mathematics." Thus was welcomed into the world the subject that soon became known as matrix mechanics.

But Born didn't let his former pupil's sarcasm deter him. Back in Göttingen he and his new assistant, Pascual Jordan, worked up a full account of Heisenberg's system in the formal language of matrix algebra. Then Heisenberg, returning after a trip to Cambridge and a restorative jaunt with his Pfadfinder brethren, joined with Born and Jordan on what became known as the *Dreimännerarbeit*—the three-man paper—which further refined and extended matrix mechanics. Heisenberg, though gratified that his physical intuition had served him well, nonetheless shared at least a twinge of his friend Pauli's skepticism. He disliked the name "matrix mechanics," thinking it too redolent of pure mathematics, of a kind that was moreover unfamiliar and off-putting to most physicists.

A simmering dispute took root here. All his life, Born nursed a resentment that his and Jordan's contributions to quantum mechanics were undervalued or even overlooked. It was "awfully clever of Heisenberg," he admitted, to have come up with matrix algebra without knowing what it was, but at the same time he seemed unable to grasp the magnitude of Heisenberg's conceptual leap. Only when he and Jordan had fleshed out the idea with necessary mathematical rigor, he believed, could it really be called a theory. That was characteristic of Born. Not a man given to physical insight, he failed to appreciate the power of scientific intuition in others. Saying that Heisenberg was "awfully clever" seems to imply that he thought his young colleague was some sort of idiot savant struck by lightning.

In any case, matrix mechanics did not win a rapturous reception from the community of physicists. They first had to learn this new branch of mathematics, then, having done so, struggled to understand what, physically, the matrices represented. Quantum mechanics, in matrix algebra disguise, was horribly complicated. At the same time, it seemed to be largely a formal achievement. The mathematical physicists claimed it was logically sound, and that it captured neatly many of the puzzling propositions that infested quantum theory. That was all very well, but what could you do with it?

Pauli's ambivalence continued. Shortly after the Born-Jordan paper appeared, he wrote to a colleague that "the immediate task is to save Heisenberg's mechanics from being drowned further in formal Göttingen scholarliness and to more clearly bring out its physical essence." Heisenberg at one point lost his cool with Pauli's scathing attitude and angrily wrote to him that "your endless griping about Copenhagen and Göttingen is an utter disgrace. Surely you will allow that we are not deliberately trying to ruin physics. If you're complaining that we're such big jackasses because we haven't come up with anything physically new, maybe you're right. But then you're just as much of a jackass, since you haven't achieved anything either."

Stung, Pauli set to work, and in less than a month had managed to use matrix mechanics in all its pure glory to derive the Balmer series of spectral lines for hydrogen—the same thing Bohr had done so many years earlier with his first simple model. Pauli's calculation was a tour de force, a powerful and convincing demonstration that matrix mechanics was more than mathematical formalism. "I hardly need tell you," a mollified Heisenberg now wrote, "how thrilled I am about the new hydrogen theory, and how amazed that you worked it out so quickly."

On the other hand, Pauli's proof was no walk in the park. The fiendish mathematics still frightened most physicists, and the claim that matrix mechanics was intellectually profound meant nothing if you couldn't follow the reasoning.

Further confusion arrived in November 1925, in the form of an elegant paper by Paul Dirac, a young physicist at Cambridge. Dirac, it appears, hadn't met Heisenberg on his recent visit to Cambridge, but saw a copy of the paper that Heisenberg dropped off. Dirac digested Heisenberg's insight and came up with his own rigorous mathematization of quantum mechanics, similar to what Born and Jordan had worked out but with a different foundation. Dirac reached back into an obscure corner of classical mechanics to find a differential operator that also obeyed the Heisenberg multiplication rule. Matrix-like elements appeared in Dirac's calculus, but in a secondary way.

It all fit together, apparently. Yet it was profoundly confusing that quantum mechanics could be dressed up as two different although evidently related systems of mathematics. In Göttingen, naturally, they liked matrices, but in Copenhagen Dirac's elegant and, as it turned out, broader and more powerful analysis won approval.

Physicists outside these select circles, meanwhile, wondered if anyone would produce a version of quantum mechanics they could understand. And that was why Schrödinger's wave equation, when it appeared in early 1926, was so gratefully received. It contained no funny algebra, just old-fashioned differential equations. Schrödinger himself made no bones about his attitude to matrix mechanics. He "was scared away," he wrote, "if not repulsed, by its transcendental algebraic methods, which seemed very difficult to me."

Sommerfeld also saw the advantages of the wave equation. Matrix mechanics, he thought, was "extremely intricate and frighteningly abstract. Schrödinger has now come to our rescue."

But Schrödinger had a larger agenda. He wanted not simply to promote an easier version of quantum mechanics but to undo some of the damage quantum mechanics had wrought. In his Nobel Prize lecture from 1933, Schrödinger talked of how, as he wrestled to create his wave equation, it had been uppermost in his mind to save "the *soul* of the old system" of mechanics.

Schrödinger insisted that a particle was not a tiny billiard ball but a tightly gathered packet of waves that created the illusion of a discrete object. Everything, fundamentally, came down to waves. There would be an underlying continuum, with no discontinuities, no discrete entities. There would be no quantum jumps, but instead smooth transformations from one state to another.

None of this followed directly from Schrödinger's equation. It was what Schrödinger hoped his wave equation would lead to. In July 1926 he lectured in Munich on his wave vision of quantum mechanics. Heisenberg was in town, having come from Copenhagen with the double purpose of visiting his parents and listening to Schrödinger in person. He admired the practical utility of wave mechanics, the way it made simple calculations possible. But he didn't like Schrödinger's broader assertions and rose from the audience to express a few objections. If physics was to be once again entirely continuous, he asked, how was it possible to explain the photoelectric effect or Compton scattering, both of which by this time amounted to direct experimental evidence for the proposition that light came in discrete, identifiable packets?

This brought an irritated response from Willy Wien, who no doubt still cherished warm memories of Heisenberg's abysmal performance at his doctoral defense just three years earlier. Jumping in before Schrödinger could speak, Wien said, as Heisenberg remembered it, "that while he understood my regrets that quantum mechanics was finished, and with it all such nonsense as quantum jumps, etc., the difficulties I had mentioned would undoubtedly be solved by Schrödinger in the very near future."

But Sommerfeld, after hearing Schrödinger, also began to have doubts. "My overall impression," he wrote to Pauli shortly afterward, "is that 'wave mechanics' is certainly an admirable micromechanics, but that it doesn't come close to solving the fundamental quantum puzzle."

Heisenberg's objection to wave mechanics was not merely technical. He didn't approve of its style. In formulating the concept behind matrix mechanics, Heisenberg had overtly put observational elements—the frequency and strength of atomic transitions—in a central role, while the undetectable motion of individual electrons remained behind the scenes. Schrödinger's waves sought to restore the older perspective. Particles, according to Schrödinger, were merely manifestations of underlying waves, but while these waves were fundamental, they were not, so it appeared, directly detectable. Wave mechanics promoted a veiled quantity to theoretical primacy, and this was not, Heisenberg profoundly believed, the right way to construct quantum mechanics.

The apparent simplicity of Schrödinger's waves was highly misleading, Heisenberg thought, and physicists were fooling themselves if they thought the Schrödinger method represented a restoration of classical values. It was not long before that suspicion was borne out.

Chapter 11

I AM INCLINED TO GIVE UP DETERMINISM

Göttingen produced matrix mechanics. Wave mechanics came from Zurich. Other voices chimed in from Copenhagen and Cambridge. From their Olympian perch in Berlin, meanwhile, Albert Einstein and Max Planck surveyed the scene. Einstein was a few years short of fifty, Planck almost seventy. Both were by now essentially conservative figures. As long as there was confusion over the apparently contradictory mathematical forms of quantum mechanics, and concomitant mystification about the physical import of the theory, both men could cling to the hope that something closer in spirit to classical thinking might yet emerge.

One aspect of the confusion dissolved with surprising ease and rapidity. In the spring of 1926 Schrödinger found that wave mechanics and matrix mechanics were not fundamentally different after all. Despite their seemingly contradictory appearances, they were in effect the same theory dressed up in strikingly different mathematics. In a nutshell, Schrödinger's waves can be used to calculate numbers that obey matrix algebra, while matrix algebra, applied to the appropriate quantities, can be made to yield Schrödinger's equation. Schrödinger was not alone in finding this remarkable equivalence. Pauli had proved it too, in a letter to Jordan, although apparently the proof didn't rise to his exacting standard of publishability, and just a little later the same argument appeared in the *Physical Review*, in a paper written by Carl Eckart, a German-American theorist from a young but promising establishment calling itself the California Institute of Technology.

But these demonstrations of the mathematical equivalence of the two versions of quantum mechanics only made it that much harder to understand how two such different portrayals of physics could arise from the same source. Physicists continued to find Schrödinger's waves comfortably familiar, while matrix mechanics remained inscrutably alien. Was there one best way to talk about the physics, or did it come down to questions of taste and convenience?

Eager to stay abreast of the developing drama, Einstein and Planck invited the principal actors to Berlin. Heisenberg came first to what he called "the chief citadel of physics in Germany," though he surely knew that in quantum mechanics the provinces flourished while the capital languished. His lecture to the distinguished Berlin professors did not seem to stick particularly in Heisenberg's mind. Far more memorable was his first searching conversation with Einstein. He had hoped to see the great man four years earlier in Leipzig, but Einstein had stayed away from that meeting after the assassination of Foreign Minister Rathenau, and Heisenberg had fled after being robbed. Back then, Heisenberg had been a mere twenty-one-year-old, still a little shy, wrestling with the dubious half-quantum business. Four years on, Einstein was still Einstein, well on his way to becoming the shaggy-maned, shabbily dressed figure of popular legend, but Heisenberg was not at all the same young man. He had held his own in disputes with Sommerfeld, Pauli, and Bohr. He had found the key to quantum mechanics. In appearance he was still the same clean-cut, unassuming figure he had always been (he looked like a peasant boy, Born had remarked on first seeing him in Göttingen; like a carpenter's apprentice, someone in Copenhagen said), but his confidence had grown. In quantum mechanics, he was the expert, Einstein the critic.

After the lecture, the two walked through the streets to Einstein's home, arguing back and forth. Einstein objected sharply to the obscurities of matrix mechanics, the way it sent position and velocity to the back of the room and brought enigmatic, unfamiliar, abstrusely mathematical quantities to the fore. Heisenberg protested that these strange developments had been forced upon him because he was trying to build a theory on what the physicist can actually observe about an atom, not on its unknown and perhaps unknowable internal dynamics. In any case, Heisenberg asked, wasn't this essentially the same strategy that Einstein had used with such stunning success years before, when he came up with special relativity?

To which Einstein could only grumble in response, as Heisenberg tells the story, that "possibly I did use that kind of reasoning . . . but it is nonsense all the same."

In devising relativity, Einstein reinvented space and time. His starting point had been to inquire closely into the meaning of simultaneity. In Newtonian mechanics, time was absolute. If two events happened in different places at the same time, then their simultaneous occurrence was an objective fact, an indisputable datum. But Einstein had the wit to ask how observers of these two events could know that they happened at the same time. They would have to synchronize their watches, as characters in war movies used to say. That meant exchanging signals—by flashes of light, by talking on the radio. But these signals travel, at most, at the speed of light, and by scrupulously following how different observers would in practice establish the times and locations of events, Einstein showed that in general they could not agree on simultaneity. Two events that happened, according to one observer, at the same time would be seen by another as happening one after the other.

In much the same way, Heisenberg insisted, it was no good imagining you could construct an absolute, God's-eye view into the inside of an atom. You could only observe in various ways the atom's behavior—the light it absorbed and emitted—and infer as best you could what was going on inside.

Einstein wasn't buying it. In relativity, although observers may disagree, events retain a distinct and unarguable physicality. A collection of observers comparing notes could arrive at a mutually acceptable consensus on what they had all seen, because special relativity accounts for the discrepancies between their individual stories. An underlying objectivity persists.

That, as Einstein saw it, was far from the case with quantum mechanics. Heisenberg appeared to be saying, he thought, that it was foolish even to ask for a consistent depiction of an atom's structure and behavior. Matrix mechanics especially, so it seemed to him, high-handedly ruled out of order questions about an electron's disposition that physicists had always given themselves the right to ask. And, Einstein firmly believed, were perfectly entitled to continue asking.

Heisenberg pushed back. Relativity had been controversial because it undermined the old questions physicists had always asked about space and time, and forced them to ask new ones. That did not mean that space and time became meaningless. He and his colleagues were trying to do the same thing for atoms—figure out the right questions to ask. Old kinds of knowledge would be lost, to be sure, but new ones would come in their place.

He had to admit, though, that he hadn't yet worked it all out. Quantum mechanics was still a work in progress. The conversation trailed off inconclusively.

Schrödinger's wave mechanics, by contrast, seemed to Einstein to offer hope. The standing wave picture of an electron in an atom had a tangible air about it. Not long after his meeting with Heisenberg, Einstein wrote to Sommerfeld that "of the recent attempts to obtain a deeper formulation of the quantum laws, I like Schrödinger's best . . . I can't help but admire the Heisenberg-Dirac theories, but to me they don't have the smell of reality."

By this time Schrödinger had also visited Berlin. Einstein found him most amiable. Schrödinger was Viennese, cultured, warm, and sophisticated. Both men were married, because they liked having someone to take care of them, but both found their pleasures elsewhere and convinced themselves their wives took this in good spirit. Though he spent years in Berlin, Einstein never felt at home among the "cool, blond Prussians." Heisenberg was by birth a southern German, from Bavaria, but his family was northern in its culture and habits. He inclined to formality and polite manners, which to Einstein came across as stiffness and reserve. Schrödinger, by contrast, was a man Einstein could feel at ease with.

But congeniality did not prevent Einstein from seeing the flaws in Schrödinger's ambitions for physics. Lecturing in Berlin, Schrödinger expanded on his hope that the waves of his equation would turn out to be direct physical pictures of electrons and other entities—not particles as such, but concentrations in space of mass and charge. Einstein was sympathetic but wary. Schrödinger was clearly expressing a hope, not a demonstrable argument. It might, Einstein could easily see, be mere wishful thinking.

Heisenberg put it more crudely. Of Schrödinger's physics, he wrote to Pauli, "the more I think about [it] the more repulsive I find it . . . to me, it's crap . . . but excuse this heresy and speak of it no more."

Schrödinger had published a brief argument in support of his interpretation. The waveform corresponding to a particle sailing through empty space, he showed, would hold together indefinitely. This physical integrity, Schrödinger argued, made the bunched-up wave an acceptable stand-in for a traditional particle.

But this result was the exception, not the rule. Max Born used wave mechanics to ponder a more complicated case, the collision of two particles, and came to a very different conclusion. After collision, he found, the waves corresponding to the rebounding particles spread out something like ripples on a pond, which by Schrödinger's interpretation would seem to mean that the particles themselves had become smeared out in all directions. That made no sense. A particle, even if it were a concentrated wave motion, must ultimately be identifiable in a classical sense. In Bohr's language, this was an instance of the correspondence principle, that the quantum description of a collision had to pass over, in the aftermath, to a suitable classical description. More fundamentally, it was just a question of common sense. A particle had to *be* somewhere; it couldn't disperse uniformly throughout space. The end result of a collision had to amount to two distinct particles moving off in well-defined directions. That's what happened in the Compton effect.

Thinking along these lines, Born came to a neat conclusion. The spreading waves leaving the collision site described, he proposed, not actual particles but their *probabilities*. In other words, a direction where the wave was strong was a direction in which rebounding particles were likely to emerge. Where the wave was weak, by contrast, particles were less likely to be seen.

If this was so, Schrödinger's equation generated not a classical wave but something wholly new. In the case of an electron in an atom, the wave must represent not some physically spread-out mass or charge, but rather the chance of finding an electron here, there, or somewhere else.

This depiction, odd as it was, harmonized with matrix mechanics. Heisenberg had defined the position of an electron in a backward manner, expressing it as a composite of the atom's electromagnetic characteristics. In a sense, Heisenberg had thereby depicted the electron's physical presence as a combination of things it might be doing, rather than some specific indication of where it was.

Born's recognition of wave mechanics as dealing with probability didn't just clarify what Schrödinger's equation meant. It also fleshed out the physical as opposed to the purely mathematical connection between wave mechanics and matrix mechanics. The price to be paid for this recognition was the intrusion into physics of probability in a new form.

Yet this conclusion slipped into the tight world of the quantum physicists with no fanfare. No one seemed to take any special note of Born's argument. That his result excited little immediate attention gave Born, in later years, further cause for bitterness. Other physicists were inclined to say in retrospect that of course they knew Schrödinger's ideas on the meaning of the waves were clearly wrong, and, yes, they could see that the waves connoted probability. Heisenberg in particular would say that the meaning of matrix elements as probabilities had been evident to him from the outset—although he didn't trouble to write this down anywhere. Textbooks on quantum mechanics, even those written soon after the subject's genesis, tended to state the definition of probability but give it no particular attribution, as if it were a step too obvious to warrant further explanation.

On the other hand, Born himself, in a later interview, acknowledged that perhaps he didn't see at the time just how revolutionary his result was. Physicists in those days all knew about the statistical physics of the nineteenth century, and many had dabbled with the idea that such statistical uncertainty might run deeper still. There was the link, first made clear by Einstein, that the intensity of emission lines from an atom had to do with the likelihood of one internal transition versus another. There had been, too, the intermittently appealing suggestion that perhaps conservation of energy would turn out to be only statistically true. As Born put it, "we were so accustomed to making statistical considerations, and to shift it one layer deeper seemed to us not very important."

Yet this later sentiment is belied by Born's own words from his 1926 paper. There he noted that it was no longer possible to say what the specific outcome of a collision would be. You could only specify the probabilities of a range of outcomes. "Here the whole problem of determinism arises," he then wrote. "[In] quantum mechanics there exists no quantity which in an individual case determines the result of a collision . . . I myself am inclined to give up determinism in the atomic world."

Determinism was the linchpin of classical physics, the crucial principle of causality. Born was now putting into words Einstein's greatest fear, one he had expressed repeatedly for years. In classical physics, when anything happens, it happens for a reason, because prior events led up to it, set the conditions for it, made it inevitable. But in quantum mechanics, apparently, things just happen one way or another, and there is no saying why.

If Born evidenced confusion about the meaning of his discovery, Einstein decidedly did not. Toward the end of 1926, he wrote to Born in words that have become famous through repetition, not least by their author, who liked his phrasing so much he would trot it out at every opportune moment. "Quantum mechanics is very imposing," he told Born. "But an inner voice tells me that it is not the real McCoy. The theory delivers a lot but hardly brings us closer to the secret of the Old One. I for one am convinced that *He* does not throw dice." If probability were to replace causality, then as far as Einstein was concerned the rational basis for constructing theories of physics had been swept away.

But younger physicists, as usual, blithely disdained such metaphysical fretting and quickly latched onto the identification of Schrödinger's waves as a measure of probability. Bohr, still the guiding spirit, approved. But others chose to bow out, notably the men who had invented wave mechanics, Louis de Broglie and Schrödinger himself. Following his precocious insight that particles must have wave properties, de Broglie duly collected a Nobel Prize in 1929 but made no further significant contributions to quantum mechanics. All his life he insisted that the probability interpretation was wrong.

Schrödinger likewise became from this time on more of a critic of quantum mechanics than a contributor to it. In September 1926, he visited Copenhagen, not long after Heisenberg had succeeded Kramers as Bohr's assistant. Bohr wanted, so he said, to hear Schrödinger's views firsthand, to understand them better.

In the event, Bohr pressed and hounded Schrödinger to explain himself from the moment he arrived, questioning his visitor in his standard relentless manner, a style that Bohr took to be the natural habit of scientific inquiry but which seemed to Schrödinger like an interrogation of Kafkaesque inescapability. Schrödinger became tired and ill and took to his bed at the institute. Mrs. Bohr fussed over him with tea and cakes while Bohr perched at the end of the bed, day and night it must have seemed, saying, "but Schrödinger, you must at least admit that . . ."

Heisenberg took only a modest part in this banging of heads. He recalled Schrödinger wistfully suggesting that some way might still be found to obtain Planck's 1900 formula for the spectrum of electromagnetic radiation without the need for quanta. "There is no hope of that," Bohr told him, speaking crisply for once. Schrödinger tried to resist, telling Bohr that "the whole idea of quantum jumps leads to nonsense," and that "if we are going to have to put up with these damn quantum jumps, I am sorry that I ever had anything to do with quantum theory," at which Bohr soothed him by saying that "the rest of us are very thankful" for wave mechanics, because of its clarity and simplicity.

There was no rapprochement. Schrödinger became angry, Heisenberg recalled, but had no answer to Bohr's soft-spoken but unending assault. Exhausted, he retreated to Zurich with his views unchanged.

Einstein, unhappy, continued to press his objections. Toward the end of 1926 he was writing to Sommerfeld that the great technical successes coming from Schrödinger's equation tended to obscure the deeper question of whether it genuinely offered a complete picture of what he quaintly insisted on calling "real events." "Are we really closer to a solution of the puzzle?" he plaintively asked.

Increasingly, Einstein spoke and wrote in the suggestive, allusive way he later became famous for. Other physicists heard more than they wanted to know about the secrets of the Old One, about the God who doesn't play dice, about the Lord being subtle but not malicious. Einstein talked as if he alone could know the inner truths of nature. His unhappiness was for this reason unanswerable. He objected to the presence of probability in physics but had found no way of getting rid of it. And the problem was about to get worse.

Chapter 12

OUR WORDS DON'T FIT

As Heisenberg and Schrödinger, along with their allies and critics, tussled over the meaning of the physics they were creating, the forty-one-year-old Niels Bohr held on to his role as guide and guru. Increasingly, though, other physicists questioned his judgments and fretted over his opaque pronouncements. Schrödinger, recovering from his ordeal in Copenhagen, confessed to frustration in dealing with Bohr. "The conversation is almost immediately driven into philosophical questions," he wrote to a friend. "Soon you no longer know whether you really take the position he is attacking, or whether you really must attack the position he is defending."

In September, Paul Dirac arrived in Copenhagen for a six-month visit. Of Bohr's famously allusive way of lecturing, Dirac observed that audiences were "pretty well spellbound," but as for himself he complained that "[Bohr's] arguments were mainly of a qualitative nature, and I was not able to really pinpoint the facts behind them. What I wanted was statements which could be expressed in terms of equations, and Bohr's work very seldom provided such statements."

Dirac, a cool, laconic loner, could hardly have been more different from the gregarious Bohr. Dirac's legendary taciturnity came about because his father, a naturalized Briton of Swiss origin, used to insist that his boy speak French at the dinner table, and, as Dirac explained later, "since I found I couldn't express myself in French, it was better for me to stay silent than to talk in English. So I became very silent at that time—that started very early." On top of that, his parents apparently had no friends, never went out, and never invited anyone to their house, so young Paul had few opportunities for small talk in English either.

Dirac respected Bohr but failed to become starry-eyed and worshipful in his presence. Perhaps for that very reason Bohr found the tall, silent Englishman strangely admirable. While Bohr struggled to put broad philosophical concepts into words, Dirac said little but sought a terse clarity in the pure logic of mathematics and unveiled his formulations—precise, if a little arid—only when he was sure of every detail. He recognized, though, that a full and systematic mathematical expression of quantum theory was not the whole story. As he said in his dry way, "getting the interpretation proved to be rather more difficult than just working out the equations."

Dirac was generally happy to play his part and leave matters of interpretation to others. That sort of laissez-faire attitude did not suit Heisenberg, who found himself increasingly at odds with his mentor Bohr. The two of them became embroiled in a tense, delicate dispute that neither man could let alone. Heisenberg had invented quantum mechanics, after all; he could hardly fail to assume some proprietorial right over the way it was portrayed and used. Bohr, on the other hand, could not altogether shake off his first impression of Heisenberg as a somewhat callow scientific thinker, piercingly imaginative but just as often wayward and impetuous. At this point in the game, Bohr thought, wisdom was required, and who was the man for that?

In Copenhagen, the two men would spend hours together during the day, Bohr talking as always in his unrelenting, insistent way while Heisenberg, animated and agitated, struggled to interrupt. In the evenings they would often continue the haggling as they took a turn around the pleasant grassy park that adjoined the institute. Often, too, even late into the night, Bohr would knock on the door of the attic room at the institute where Heisenberg was staying, offering just a small clarification or emendation of what he had been trying to say earlier. Not infrequently these footnotes to the day's discussion would sprawl into the small hours. Bohr would stick to no fixed schedule. Whatever had to be said had to be said there and then. Clashing like this for weeks, both men grew weary of the argument, and of each other.

What they argued over during these never-ending days in late 1926 was, in one form or another, the question of continuity versus abruptness. Schrödinger, of course, wanted it all to come down to waves, with discrete particles and their capricious behavior merely an illusion. That, Heisenberg and Bohr could at least agree, was a lost cause. But Heisenberg, having enthusiastically jettisoned the old ways, characteristically wanted to run to the opposite extreme and embrace the most radical thinking at all costs. Quantum mechanics forced physicists to think in new ways, to learn a new language. Too bad, said Heisenberg. They would have to get used to it.

To Bohr, that attitude was cavalier—or, what was worse, shallow. As he pointed out repeatedly and forcefully, position and velocity and all the other old reliables of classical mechanics had not suddenly lost all their utility. In the world outside the atom, the old concepts continued to serve well. There had to be, Bohr insisted, a connecting up. You had to be able to get from the discontinuity and discreteness of the quantum world to the smooth continuity of the familiar classical world.

Heisenberg found Bohr's attitude frustrating, almost deliberately so, as if frustration were a desirable state, a commendable aspiration. It was as if Bohr wanted to find a way to talk about quantum mechanics in classical language while at the same time freely admitting it couldn't be done—not, at least, without contradiction and inconsistency. But Bohr positively reveled in contradiction; it constituted his own internal Socratic discourse.

Whenever Heisenberg claimed he understood how quantum mechanics worked, or at least that he could reliably make use of it, Bohr would just as reliably find an obscure point, a lack of logical clarity. "Sometimes," Heisenberg recalled, "I had the impression that Bohr really tried to lead me onto *Glatteis*, onto slippery ground . . . I remember that sometimes I was a bit angry about it." But he ruefully acknowledged that if Bohr could so dependably put his finger on subtle problems, then perhaps they were on slippery ground after all.

This banging of heads could go on only so long. By early 1927, Bohr and Heisenberg had stated and restated their opinions so often that they were talking past each other, reduced to helpless frustration that neither one could or would acknowledge what the other was saying. In February, Bohr went to Norway, to spend some time skiing. Originally he had planned this as a trip for the two of them, but now it seemed better to go alone. Heisenberg, meanwhile, could trudge by himself around the park in the early evening, without Bohr dogging his every step.

But the nagging echo of Bohr's voice stayed with him. Suppose it was true, as Bohr asserted, that position and velocity must continue to have meaning, even if it was not the traditional meaning physicists had always assumed. What would that new meaning be? How could he get at it?

In their wrangling thus far, Heisenberg and Bohr had treated the issue as a theoretical one. Classical mechanics worked with one set of precepts, quantum mechanics with another, and how were the two to be reconciled? It was, to borrow Dirac's phrase, a matter of getting the interpretation, of hearing what the mathematics was trying to say. Dirac, in fact, had provided an important clue, though Heisenberg hadn't immediately latched onto it.

While in Copenhagen, Dirac had put the finishing touches to his magisterial presentation of quantum mechanics, in which he showed in a perfectly general way how to take some problem in classical mechanics and define its quantum equivalent. He could also do the reverse—that is, he could show how some quantum mechanical system would look if you insisted on describing it in classical terms. But in that translation, he found, a curious discrepancy arose. Beginning with some quantum system of particles, for example, you could work out a classical picture in which the positions of the particles were the primary elements, or you could choose instead to speak in terms of particle velocities—or rather, in terms of momentum, mass times velocity, which to physicists is the more fundamental quantity. Strangely, though, these position and momentum portraits didn't match up as they should, if they were merely alternative portrayals of a single underlying system. It was as if the position-based account and the momentum-based account were somehow depicting two different quantum systems, not the same one in different ways.

Pauli had come across the same awkwardness. He wrote to Heisenberg about it, using p as a standard notation for momentum, while the adjacent letter q then stood for position. "You can look at the world with the p-eye," he said, "and you can look at it with the q-eye, but if you want to open both eyes at the same time, you will go crazy."

Quantum particles wouldn't reveal themselves clearly. They yielded up contradictory pictures. That was the conundrum Heisenberg wrestled with. How could he find a way to force quantum mechanics to give up its secrets, to let him see what was going on inside?

He couldn't! That was the answer that flashed into his mind one evening as he plodded around the park, lost in thought. Just as, on Helgoland, he had realized that it would never be possible to describe quantum jumps in the continuous language of classical physics, so now the same lesson bore in on him, in a yet larger way. There was no way to force a quantum system to yield up a description that would make unambiguous sense in classical terms.

Well, yes, but wasn't that what he had been trying to tell Bohr for months now? Except now he began to see Bohr's point of view. You might not be able to come up with an unambiguous account—but that didn't mean, as Heisenberg had thought until now, that you just gave up trying and moved on. You had to find some way of talking about quantum systems.

At last Heisenberg was able to grasp a point that neither he nor Bohr had understood so far. The crucial question was not a theoretical one, still less, as Bohr often seemed to think, a philosophical one. It was in the end a *practical* matter.

You might not be able to talk about the position and momentum of quantum objects in a way that would make sense under the old rules. But what you can still do, Heisenberg now saw, is what physicists have always done—you can attach meaning to position and momentum by *measuring* them. The way to cut through theoretical confusion was to pay attention to practicalities.

He just needed to think of a simple example to make his insight plain. And so, perhaps with Compton's pretty experiment of a few years earlier in the back of his mind, he hit upon the disarmingly straightforward example that has made his name iconic. An electron flies through space. An observer shines light upon it, then detects the light that bounces off the speeding particle. By measuring this scattered light—its frequency and direction—the observer can try to deduce the position and momentum of the electron at the moment the light hit it. And that, as Heisenberg discovered, is where things get interesting.

Light consists of quanta—or photons, as they had recently been dubbed by the American physical chemist Gilbert Lewis. The encounter between one of these photons and the flying electron is a quantum event. That encounter, as Born had proved, doesn't yield definite outcomes, but a range of possible outcomes, with various probabilities. Reversing the logic, Heisenberg now realized that an observer cannot infer a single unique event that would have led to the measured outcome. Instead, a range of possible electron-photon encounters could have happened. Which must mean, he saw, that it would be impossible to infer uniquely what the position and momentum of the electron was.

Pauli had said you could look at position or you could look at momentum, but you can't look at both at once. Heisenberg, thinking the matter carefully through, realized it wasn't as simple as that. It wasn't either-or, but an inescapable compromise. The more an observer tried to extract information about the electron's position, the less it was possible to know about its momentum, and vice versa. There would always be, as Heisenberg put it, an "inexactness" (*Ungenauigkeit*) in the conclusions.

It was during Bohr's absence that Heisenberg persuaded himself of this tidy but startling result. He had learned to be wary of Bohr's intense scrutiny of new ideas. He wrote Pauli a long letter explaining what he had come up with, but to Bohr he sent only a brief note saying that an interesting development awaited his return. By the time Bohr got back, Heisenberg had already sent his paper off for publication. Bohr read it, grew fascinated, then gravely troubled.

Heisenberg described an encounter between two particles, a photon and an electron, and found an inexactness that derived from the unpredictability of that collision. Bohr—inevitably, exasperatingly—came up with another way to look at the matter. An observer detecting the photon measures it not as a particle but as a little bundle of waves. And in classical optics, he reminded Heisenberg, waves have limited resolving power. That is, light of a certain wavelength cannot render clear images of any object smaller than that wavelength. The picture becomes blurry. And *that*, Bohr said, was the explanation for what Heisenberg had found. It was in the act of using information from a wave measurement to infer the properties of a particle that inexactness sprang up.

Bohr's reinterpretation outraged Heisenberg. First, because Bohr was dragging waves back into it, which bore the taint of Schrödinger's name, and second, because Bohr's argument seemed to be about the limitations of classical optics, not the unpredictability of quantum events.

But no, Bohr retorted, that wasn't it either. It was precisely because of the mixing of incommensurable concepts—particles and waves, quantum collisions and optical resolving power—that inexactness crept in. It was the outward sign of the internal mis-match between quantum and classical principles. This interpre-tation, as it happened, fell in very neatly with ideas that Bohr had been pondering while he skied alone in Norway. He had evolved a broad new principle, soon to be christened "complementarity," according to which both the wave aspect and the particle aspect of quantum objects had necessary but contradictory roles to play. Depending on the problem, one aspect or another might come to the fore, but neither could be neglected entirely. Heisenberg's inexactness, he declared, was the demonstrable evidence of this unavoidable disharmony.

Heisenberg was aghast. He had worked out an elegant result in a straightforward way. Now Bohr wanted to smother it in the thick metaphorical garb that he loved but that Heisenberg found so oppressive. Heisenberg wanted to go ahead and publish his discovery. Bohr wanted him to contact the journal and hold the paper up while they worked out together the best presentation of the physics. Heisenberg refused. Bohr then found a technical er-ror in Heisenberg's analysis that was, to Heisenberg's enormous chagrin, reminiscent of the error he made years ago in his thesis defense as he had tried to answer Willy Wien's questions about standard optical theory. Heisenberg insisted it wasn't a big prob-lem and pushed on. Eventually, in May, he reluctantly agreed to add an endnote to his paper, just before it went to press, thanking Bohr for clarifications and allowing that the precise source of ob-servational "uncertainty"—he now used that word, which Bohr preferred—was perhaps not as evident as the author's presenta-tion implied.

It was in this painful, quarrelsome way that Heisenberg's famous uncertainty principle entered the world. As Bohr and Heisenberg wrestled back and forth over how best to express it, the unavoidable difficulty, said Heisenberg, was that "our words don't fit."

Certain words caused particular difficulty. Writing wearily to Pauli, Heisenberg remarked that "all the results in the paper are certainly correct and Bohr and I are in agreement about them—only between Bohr and myself there are considerable differences in taste over the word '*anschaulich*.' " This adjective has caused problems for German-speaking physicists, still more for those faced with translating it into English. Heisenberg titled his paper on inexactness "*Über den anschaulichen Inhalt der quantentheoretischen Kinematik und Mechanik,*" which has been rendered by one author as "On the *Perceptual* Content of Quantum Theoretical Kinematics and Mechanics," by another as "On the *Physical* Content . . . ," while a third translates *anschaulich* as "intuitive." It is as if a single word can mean both "concrete" and "abstract."

The verb *anschauen* means "to look at"; something *anschaulich* is therefore something capable of being looked at. Heisenberg means to speak about phenomena that the physicist can in principle observe, hence the translation of *anschaulich* as "perceptual"—that is, perceivable. Hence too its rendering as "physical"—meaning quantities that are empirically meaningful in the traditional way. And thence, with a hop and jump, comes "intuitive," because the quantities that make sense to physicists are those, such as position and momentum, that have familiar or commonsense meaning. (The flaw here is that no one thought of momentum as intuitive until Newton invented it and made it part of every later scientist's common sense.)

Equally tricky is the more famous word that came into physics and thence into wider circulation. In talking about experimental measurements, Heisenberg consistently used the word *Ungenauigkeit,* "inexactness." But in one section of his paper, referring to the theoretical point both Dirac and Pauli had made about ambiguity in the theoretical description of a system, he switched to *Unbestimmtheit*, from the verb *bestimmen*, "to determine." He made a distinction, that is, between the inexactness of experimental outcomes and the indeterminacy of mathematical descriptions. Only in his endnote does there appear abruptly the word *Unsicherheit*, "uncertainty," which was Bohr's choice and which through Bohr made its way into the vocabulary of English-speaking physicists.

"Inexactness," in truth, is a poor word to describe what Heisenberg found, since it fails to distinguish the new inability he pinpointed from the ubiquitous and long-standing difficulty of making any measurement exactly. A few old-fashioned physicists still prefer to speak, in English, of the indeterminacy principle, which is a better way of putting it. (In the afterword to his play *Copenhagen*, Michael Frayn suggests, a little more pointedly still, "indeterminability.") German-speaking physicists today refer to *die Unschärfe Relation*, a nice choice. In German as in English, sharpness is the quality of a well-made photograph, so *unscharf* means "blurred." To speak of the blurriness principle suggests a pleasant connotation, that the more you squint and peer, the less you can make out whatever it is you are trying to see. But "blurriness" is no doubt an insufficiently grand word to enter the English scientific lexicon at this late hour.

"Our words don't fit," Heisenberg told Bohr, and perhaps he switched from one word to another because he reckoned no word would perfectly capture his idea. But Bohr seemed to think he could find the right words or phrases, if only he kept trying. Only by expressing quantum mechanics in familiar terms, he insisted, could physicists hope to make sense of it as something more than a set of mathematical relationships.

In June 1927, Pauli visited Copenhagen, hoping to act as a

mediator between the warring principals. Heisenberg had been driven to tears at one time by Bohr's unceasing interrogation. On other occasions his frustration caused him to snap back harshly and angrily. Bohr, in all this, as in his earlier encounter with Schrödinger, seems to have maintained a serene, insufferable calm. Pauli soothed Heisenberg a little, but the dispute saw no tidy conclusion.

Heisenberg, in any case, was about to leave Copenhagen in order to take up a professional chair at the University of Leipzig. There, away from Bohr's vexing presence, Heisenberg reflected on the previous few months, and after a time wrote ruefully to Bohr, regretting how ungrateful he must have seemed. A brief visit to Copenhagen later in the year helped mend fences.

Never again, though, would these two have so close, difficult, or intense an intellectual engagement as they had during Heisenberg's time as Bohr's assistant. Heisenberg, still only twenty-six, had achieved security as professor in his own right, which among other things at last assuaged his father's frequent concern that he was squandering his intellectual talents on frivolous matters. Meanwhile Bohr, perhaps a little put out that Heisenberg, working alone, had hit upon a daring and perplexing new argument that seemed to threaten principles physicists had long cherished, took as his next task the formulation of a sound philosophy for understanding this strange concept of *uncertainty*.

Chapter 13

AWFUL BOHR INCANTATION TERMINOLOGY

For all its subsequent notoriety, the uncertainty principle's arrival did not trigger instant unrest and rioting in the halls of physics and philosophy. Born, recognizing Schrödinger's waves as representations of probability, had already said that determinism must go. Pauli and Dirac had seen that there was something strange about the way quantum physics manifested itself to the outside world. Heisenberg's uncertainty pinned down that strangeness, put a number on it, and—perhaps most important to Heisenberg—dashed any lingering hope that Schrödinger with his waves could restore some sort of classical reality to physics.

But this discussion, to the select few engaged in it, concerned the inner workings of quantum mechanics. It was Bohr, developing his new philosophy of complementarity, who grappled overtly with the way that the phenomena of quantum mechanics must make themselves known in a broader context. Complementarity, for Bohr, flowed from his idea of correspondence, that the quantum world must transform seamlessly into the classical world, which is what we continue to see all around us. Complementarity was supposed to make quantum mechanics comprehensible and practical to the great mass of working physicists. It was in this attempt at translation that the truly revolutionary aspects of quantum physics burst onto a larger stage.

After Heisenberg left Copenhagen for Leipzig, Bohr began the slow and painful process of composing his own interpretation of the uncertainty principle. With his new assistant, Oskar Klein, as amanuensis, Bohr thought out loud, practiced his pronouncements, then each morning discarded what Klein had struggled to write down the previous day, and started over. When the Bohrs went to their country cottage on the Danish coast, north of Copenhagen, for the summer, Klein tagged along. The painfully slow composition continued. Margrethe Bohr, normally cheerful and stoic, was reduced on occasion to tears—not, as Heisenberg had been, because she disputed her husband's angle on physics, but because he had gone on an extended mental absence from what should have been their family vacation. The Bohrs, by this time, had a lively collection of five children, all boys; a sixth boy would arrive the following year.

For all his dithering over phraseology, Bohr never wavered from his underlying conviction. Any practical description of a quantum object's properties or behavior must ultimately be couched in classical terms. That was unarguable. The result of any experiment was necessarily a concrete datum, not a cloud of probabilities.

Uncertainty and complementarity, Bohr thought, shed light on why Schrödinger's waves were by no means the classical constructs their author had hoped for. Formally, Schrödinger's equation is deterministic in the old-fashioned sense. That is, if you know the wave function for some system at a certain time, you can calculate it exactly and unambiguously at any later time — provided, that is, you don't attempt any observation in the interim. Measurement is what causes Born's probability interpretation of the wave to swing into action: different results are possible, with different likelihoods.

Heisenberg's uncertainty nailed down the inescapability of the discord between one possible measurement and another. An observer can choose to measure this, that, or the other, but has to put up with resulting incommensurabilities. And that uncertainty feeds into the future development of the system. The quantum wave function changes to reflect the fact that one particular measurement outcome occurred and other possibilities didn't — and that in turn influences the possible outcomes of subsequent measurements that might be made. Complementarity was Bohr's way of trying to keep all these conflicting possibilities under one roof.

Bohr presented his overarching philosophy in September 1927, at a meeting in Como, northern Italy, to mark the centenary of the death of Alessandro Volta, the Italian pioneer of electricity. Historically, his lecture there marks as well as anything the formal introduction into science of the idea that measurements are not passive accountings of an objective world but active interactions in which the thing measured and the way it is measured contribute inseparably to the outcome. At the time, though, Bohr's tortured and tortuous remarks mostly fell flat. Those who were not utterly baffled felt that Bohr was for some reason trying to say what they already knew, only in a needlessly mysterious way.

Bohr prepared an account of his Como talk for the scientific journal *Nature*. The process consumed many months in agonized redrafting, entreaties from the editor, assistance from Pauli, abject apologies from Bohr, followed by further delays. The result, printed finally in April of the following year, was accompanied by an editorial comment lamenting that Bohr had destroyed any last possibility that classical principles of physics might be restored, but hoping, by way of a poor substitute, that Bohr's elusive phrases were not "the last word on the subject, and that [physicists] may yet be successful in expressing the quantum postulate in picturesque form."

Bohr said, for example, that he and his colleagues were "adapting our modes of perception borrowed from the sensations to the gradually deepening knowledge of the laws of Nature," a statement that only with difficulty passes grammatical scrutiny and no doubt caused the average reader of *Nature* and interpreter of nature to gape helplessly.

The idea emerged, so often said and so little understood, that measurement disturbs the system being measured. But as Bohr tried to explain, *all* measurements amount to disturbances of what's being measured. The new thing about quantum mechanics, he wanted to get across, is really that measurement *defines* what is being measured. What you get from a measurement depends on what you choose to measure, which is nothing new, but as Heisenberg had now proved, measuring one aspect of a system closes the door on what else you can find out, and thus fatally restricts the information that any future measurement might yield.

At Como, Born stood up to say briefly that he agreed, mostly, with Bohr. Crucially, so did Heisenberg. Only one or two insiders knew of his and Bohr's fierce, tense standoff during the previous months. Now, apparently, all that was done with, and Heisenberg had nothing but praise and thanks for his mentor.

So it was that the so-called Copenhagen interpretation of quantum mechanics began to gain traction, a circumstance that has vexed not only physicists but also historians and sociologists of science. It has the look of a conspiracy. Despite much internal disagreement, it seems, the Bohr camp publicly united in order to squelch criticism from those not in the inner circle. Heisenberg in particular swallowed his objections, wiped away his tears, and obediently toed the party line.

Did Heisenberg, as had happened with Kramers, succumb to Bohr's irresistible force, or collapse in the face of Bohr's inexhaustible capacity to argue? Or, as some have suggested, did Heisenberg's fervent desire to win a professorial appointment in Germany necessitate an abdication to Bohr's views, in order to show that he was a solid, reliable fellow, a team player, not a hothead or a maverick?

Neither speculation seems likely. Heisenberg had, after all, shown enough resilience to insist on publishing his uncertainty paper before Bohr had properly blessed it. At the age of twenty-six, he was responsible for the essential insight that created quantum mechanics in the first place and had now worked out one of its most disturbing and far-reaching consequences. Despite fundamental disagreements about physics, he had won the admiration of Einstein and Planck. It is hard to imagine that he felt the need to suppress his own views in order to get a job.

The simple explanation is not necessarily wrong. After he left Copenhagen, Heisenberg looked back on his behavior and could see that some part of his hostility was little more than amour propre, an unhappiness that Bohr did not see uncertainty exactly as he did. Pauli chided him to give Bohr's ideas more consideration. The sweeping generality and concomitant vagueness of complementarity may not have been to Heisenberg's taste, but when it came to explaining how physicists were to make sense of quantum mechanics, he could not deny that Bohr's strategy captured an important truth. And, simply put, was helpful.

Heisenberg changed his mind, in short, because he saw that Bohr offered a better way forward. He was a pragmatist. There is no reason to believe he was insincere.

If the Como meeting was distinctly unmomentous, one reason was the absence of both Einstein and Schrödinger. Sometime in the spring of 1927, Einstein submitted a paper arguing for a realistic—that is, not probabilistic—interpretation of Schrödinger's waves, only to withdraw it, apparently after corresponding with Heisenberg. He did not like uncertainty, but his attempts to find a counterargument went nowhere. Fretting but frustrated, Einstein remained in Berlin, where Schrödinger would soon join him as a faculty colleague. Planck had officially retired, and the convivial and scientifically conservative Schrödinger came through as the most agreeable replacement.

On the face of it, Einstein ought to have liked complementarity. As early as 1909, when he alone was arguing for the reality of photons, he had said that theoretical physics must "bring us a new theory of light that can be interpreted as a kind of fusion of the wave and the emission [that is, photon] theory." And just before Heisenberg unleashed uncertainty, Einstein lectured in Berlin on the need for a synthesis of conflicting views. But such a synthesis, to Einstein, would of necessity make the underlying conflicts go away. Bohr's complementarity, by contrast, like its creator, seemed positively to revel in contradiction.

Only a few weeks after the Como meeting broke up, many of the same physicists reconvened in Brussels for the Fifth Solvay

Conference on Physics, taking as their theme "Electrons and Photons." Ernest Solvay was a Belgian chemist and amateur enthusiast of science who had made a fortune on the strength of an industrial process for the manufacture of sodium carbonate. In 1911, intrigued by the emerging physics of atoms and radiation, he had funded an invitation-only meeting at the luxurious Hôtel Métropole in Brussels, where twenty luminaries—including Einstein, Planck, Rutherford, and Mme Curie—could debate in leisurely comfort their most pressing questions.

So well received was this debut that Solvay decided to make his conference a triennial occurrence. The war interrupted that schedule, but resuming again afterward, the Solvay conferences became the venue for many of the knottiest and most profound scientific discussions of the postwar years. They remained invitation-only events, with no more than twenty or thirty distinguished participants.

Because of the postwar exclusion of German scientists, it was not until the fifth Solvay meeting of 1927 that a truly representative international assembly gathered again. Einstein returned, and Bohr came for the first time, having missed the 1924 meeting because of illness. (Solvay himself had died in 1922.) And at the fifth Solvay there was something large to discuss: quantum mechanics and the uncertainty principle, neither of which had existed three years earlier.

A striking division into old guard and young Turks emerged, except that Bohr, characteristically, refused to sit quietly in either camp. The young men, notably Heisenberg, Pauli, and Dirac, wanted nothing except to push quantum mechanics forward by applying it to unsolved problems concerning atoms, photons, and radiation. They were impatient of anything that smacked of philosophy, semantics, or pedantry. On the other side, de Broglie tried to restrain the avant-garde by talking up Schrödinger's invention of a scientifically acceptable form of quantum mechanics, while Schrödinger offered a rather inarticulate defense of his own conception of quantum waves, refusing the probability interpretation. His talk drew sharp criticism from Born and Heisenberg in particular, and Schrödinger kept his head down for the remainder of the meeting.

Einstein, answering as always to no one's vision but his own, yet standing also as the chief among traditionalists, gave no formal presentation. He had been invited to speak on his own views of quantum mechanics, but after some hesitation begged off, saying that he hadn't pondered the matter as thoroughly as he would like and preferred to sit and listen. During the conference talks he mainly bit his tongue, and kept his worries to himself. When he did occasionally rise to speak, he did so apologetically, admitting that perhaps he had not looked closely enough into quantum mechanics to be sure of what he was saying.

But Einstein made his presence felt nonetheless. Over meals, after hours, long into the evenings, he pushed the advocates of quantum mechanics to say precisely what they believed, and pressed upon them his own reservations—intuitive, philosophical, not wholly rational, but weighty all the same. There was no shortage of miscommunication, with advocates of one point of view failing to digest the objections from other camps. At some point Paul Ehrenfest, one of Boltzmann's last students and a close friend of Einstein's, inscribed on a blackboard the verse from Genesis about Babel: "The Lord did there confound the language of all the earth." Heisenberg and Pauli professed to be unconcerned with the old man's grumblings. They listened deferentially, said little, but could be heard muttering to themselves that there was nothing to worry about, it would all turn out fine.

Bohr, on the other hand, both out of personal respect for Einstein and because he too indulged in philosophical worries, could not ignore his old friend's objections. It was he who took on the task of defending quantum mechanics, as if the others did not really see that it was in any great need of defense. And Bohr admitted in private that he did not entirely understand what it was that Einstein so strongly objected to.

Einstein put up one of his favorite devices, a thought experiment. He asked his colleagues to imagine something quite simple. Think of a beam of electrons passing through a tiny hole in an opaque screen, he said. Because the electrons have wave characteristics, they will create, on a second screen placed beyond the first to record an image, a so-called diffraction pattern of alternately light and dark rings. (This phenomenon, predicted for light by the French scientist Augustin Fresnel in the early nineteenth century, had been one of the clinching pieces of evidence in favor of the wave theory of light.)

Quantum mechanics, supposedly, could predict only the probability that each electron would hit the screen in one place or another. Individual electrons passing through the hole and distributing themselves in probabilistic fashion would dutifully but independently build up the required diffraction pattern. But think about a single electron, Einstein urged. As soon as it hits the screen in one place, the probability of it hitting anywhere else must fall to zero. The wave function must abruptly change to register the new situation. Does this not imply, Einstein argued, that something instantaneous happened across the screen at the moment of impact?

Here was the germ of what became Einstein's perennial objection to quantum mechanics. It implied faster-than-light communication, though admittedly just what was being communicated was hard to fathom. Unfortunately, the only substantial account of the tussle between Einstein and Bohr was written by Bohr himself, some twenty years later. In it, we get a somewhat perplexed glimpse of Einstein's argument, followed by Bohr's detailed response, which misses the fundamental point.

Since Einstein could not countenance faster-than-light phenomena, he insisted (said Bohr) that quantum mechanics could not be the whole story. There must be some way, within a theory grander than mere quantum mechanics, of calculating the behavior of electrons in detail so that you could predict exactly where each and every one would end up. In that case, the probability inherent in quantum mechanics would turn out to be like the probability enshrined in the old kinetic theory of heat. There, atoms have definite properties at all times and behave, in theory, with absolute predictability. But the physicist cannot hope to know precisely what every atom is doing, so is forced to resort to a statistical description. Quantum mechanics ought to work the same way, Einstein insisted. Beneath the surface it ought to be deterministic in the traditional way. And the intrusion of probability would not indicate a fundamental breakdown in physical determinism, only that physicists had not yet figured out the complete picture.

By way of counterargument, Bohr used the newly minted uncertainty principle to prove that there was no way to extract more information about the electrons in Einstein's thought experiment—without, that is, destroying the diffraction pattern in the process. You could get details of each electron's trajectory before it hit the screen, or you could get the diffraction pattern, but you couldn't get both.

It's not hard to imagine Einstein's exasperation at this response. Of course quantum mechanics can't give you all the information you would like. That was precisely the problem Einstein wanted to bring into the open. Far from demolishing the difficulty, Bohr had reinforced it. Quantum mechanics couldn't be the whole story.

A letter written by Ehrenfest shortly after the Solvay meeting conveys the affair in an enthusiastic, telegraphic style. "Like a chess match," he reported. "Einstein ever ready with new arguments. Bohr always producing out of a cloud of philosophical smoke the tools for destroying one example after another. Einstein like a jack-in-the-box, every morning jumping up afresh. Oh, it was priceless." Ehrenfest was chagrined to see Einstein speaking irrationally of quantum mechanics, the way *his* critics spoke of relativity, and he said so to Einstein's face. But then he also acknowledged that Einstein's dissatisfaction made him uneasy too. And although he sided with Bohr, he couldn't resist complaining about the "awful Bohr incantation terminology. Impossible for anyone else to summarize."

Other participants did not recall the meeting in such melodramatic terms. Dirac, whose views were very much in sympathy with Einstein's, remarked coolly that "I listened to their arguments, but I did not join in them, essentially because I was not very much interested. I was more interested in getting the correct equations." Complementarity, he said elsewhere, "doesn't provide you with any equations which you didn't have before."

Their encounter at the fifth Solvay meeting brought happiness neither to Einstein nor to Bohr. Neither had usefully communicated his perspective to the other. Heisenberg and Pauli stood mostly to one side. Much later Heisenberg claimed that the Solvay meeting was important for establishing a consensus view of quantum mechanics, although when pressed he admitted that the consensus consisted of Bohr, Pauli, and himself. Lecturing in Chicago in 1929, he talked admiringly of Bohr's influence and of *der Kopenhagener Geist*—the Copenhagen spirit. For both adherents and dissenters, the Copenhagen interpretation was crystallizing into the standard view of quantum mechanics. It has been over the decades as elusive as it has been influential. Those who subscribe to it talk of its profundity and power while acknowledging they can't easily put it into words. Precisely the problem, say its critics. It has acquired de facto authority even though no one seems to be able to say quite what it is.

Einstein was not mollified. A year after the fifth Solvay meeting he wrote scornfully but resignedly to Schrödinger that "the soothing Heisenberg-Bohr philosophy—or religion?—is so nicely contrived that for now it offers the true believer a soft pillow from which he is not easily rousted. So let him lie." It is ironic, of course, that Einstein objected to religious principles in others when his authority for disliking quantum mechanics derived from his direct access to the thoughts of "the Old One."

Chapter 14

NOW THE GAME WAS WON

Toward the end of the summer of 1928, a young Russian who had just finished a two-month summer school in Göttingen stopped off for a day in Copenhagen, hoping to meet Niels Bohr before he returned to Leningrad. Finding some free time in the afternoon, Bohr listened keenly as the gangling young man, George Gamow, explained how he had worked out an elegant but odd answer to a long-standing puzzle. Bohr asked Gamow how long he intended to stay in Copenhagen. Gamow replied that he had to leave that very day, as the modest amount of money the Soviet authorities had supplied for his trip had run out. If he could arrange for a year's fellowship at the institute, Bohr asked, would Gamow be willing to stay? Gamow paused, gulped, and said yes.

What grabbed Bohr's attention was Gamow's explanation of

the old puzzle about radioactive decay, the enigma that Marie Curie had remarked on as long ago as 1898 and that Rutherford and Soddy had demonstrated quantitatively in 1902. Decay, they all saw, followed a truly random course: any unstable nucleus has a constant probability, in a given time, of disintegrating. Although this had been the first instance in physics of a truly unpredictable phenomenon, its significance had not immediately leaped to physicists' attention. Even in 1916, when Einstein remarked that electron jumps in the Bohr atom also followed the same law of probability, physicists did not entirely grasp that a new and awkward phenomenon had entered the theoretical arena. Still less could they perceive the source of any connection between radioactivity and electron jumps.

When Gamow met Bohr, little was understood of nuclear physics. The proton was known, and there was a growing belief that it must have a neutral partner. But it was 1932 before the discovery of the neutron confirmed that suspicion. Physicists had no idea what kept a nucleus in one piece: electrostatic repulsion ought to cause a tightly packed collection of positively charged protons, with or without neutral companions, to fly instantly and forcefully apart.

Necessarily, Gamow could conjure up only an exceedingly simple model of alpha radioactivity. He imagined that alpha particles, known to be identical to the nuclei of helium atoms, pre-existed inside heavy, unstable nuclei, and he assumed that whatever force held the nucleus together also stopped these alphas, most of the time, from popping out. Looking at this picture with a quantum eye, he came to a conclusion both surprising and satisfying.

Classically, a force strong enough to retain alphas within the nucleus will keep them tucked inside forever. Think of a marble rolling around inside a shallow bowl. If it has enough energy to fly over the bowl's rim, it will do so promptly, but if it doesn't have enough speed to get to the edge, it can never get out. There's a clear demarcation between the two cases.

But Gamow used Schrödinger's equation to depict one of the alphas inside a nucleus as a quantum wave rather than an old-fashioned particle. For mathematical reasons, he found, this wave could not vanish abruptly at the boundary of the nucleus. It had to extend beyond, trailing off into the distance. But if the wave existed outside the nucleus, Gamow realized, then there must be some measurable probability that the particle could actually *be* outside the nucleus. According to Gamow's quantum analysis, an alpha particle can't exist strictly and only within the nucleus.

In other words, an alpha particle has some fixed and constant probability of showing up outside the nucleus—and once it does, electrostatic repulsion will take over and send it zooming away. Gamow's simpleminded model not only gave a reason why alpha decay occurs but explained too the probability law that Rutherford and Soddy had found a quarter of a century earlier.

By the time he got to Copenhagen, Gamow had already sent off a paper for publication. As it happened, two American physicists, Edward Condon and Ronald Gurney, independently came up with the same idea and published their work too in 1928.

This model of alpha decay is usually cited as the first example of a general quantum phenomenon known as tunneling: the alpha particle can sneak through what is in classical terms an impenetrable barrier created by the confining force. But "tunneling" is an awkward attempt to translate a classically impossible phenomenon into familiar language. It suggests an image of a particle rolling around in its prison until, spontaneously, it slips through the wall and gets clear away. But in pure quantum terms—consistent with either Schrödinger's waves or Heisenberg's uncertainty—the alpha never has the definite position or momentum this classical image implies. Instead, it maintains a sort of constant but fractional existence beyond the boundaries of the nucleus.

And that raises a tricky question. If the alpha particle has at all times a certain probability of existing beyond the nucleus, why in fact does it slip away at one moment and not another?

"How does an electron decide?" Rutherford had asked Bohr all those years ago, failing to see why it would jump to a new orbit at one moment rather than another. And now the same question arose in alpha decay. How does the nucleus decide when to split?

Gamow's account of alpha decay shows that the answer to these two questions is the same. Or rather, the lack of an answer is for the same reason. Quantum mechanics gives only probabilities. That's all. To ask for a specific prediction of when or where something will happen is to ask for more than quantum mechanics can give. Classically, when something happens, there has to be an immediate cause. In quantum mechanics, that time-honored, seemingly obvious rule no longer applies. It's not hard to see why Einstein saw this as an admission of defeat rather than a scientific explanation.

Gamow was twenty-four at this time, a recent graduate of the University of Leningrad. He had missed, by just a couple of years, the heroic days of quantum mechanics, when Heisenberg, Schrödinger, Dirac, and the rest, under Bohr's watchful guidance and Einstein's skeptical gaze, had assembled the new physics. To Gamow, as to all the other physicists of the latest generation, quantum mechanics offered a marvelous set of tools with which they could tackle all manner of previously unthinkable questions. Not just nuclear physics but the physics of crystals and metals, of the conduction of heat and electricity, of transparency to light versus opacity—all began to yield to quantum mechanical insights. With a vast range of practical problems opening up to them, physicists were not inclined to dawdle over philosophical concerns. There was too much to do, and it was all too much fun.

But Einstein, never one to bother with detailed calculations of intricate phenomena, could not abandon his deep worries. He still had some fight left in him.

"At the next meeting with Einstein at the Solvay conference in 1930," Bohr recalled in what was for him uncharacteristically vigorous language, "our discussions took quite a dramatic turn." As before, thirty of the world's leading physicists assembled in Brussels, their official topic on this occasion being magnetism. The official topic and formal proceedings have largely faded from the history books. What lingers is the memory of another tense, fervent showdown between Einstein and Bohr.

Following the previous inconclusive Solvay debate, Einstein had no doubt realized that metaphysical scruples would get him nowhere. He needed a specific, quantitative demonstration that something was amiss, and by the time he arrived in Brussels, he thought he had one. He intended to prove to Bohr and his disciples that the uncertainty principle, now hailed as a fundamental principle of quantum mechanics, could not be the final truth. He had found a way around it, a way to get more information out of an experiment than Heisenberg's rule would allow.

The experiment, of course, was not a real one but another example of that favorite Einsteinian device, the thought experiment. It was a test that could not by any stretch of the imagination be done in anyone's laboratory, but it was one that the laws of physics permitted. More to the point, according to Einstein, the laws of physics in this case proved that the experiment would yield better results than Heisenberg would allow. It was so simple as to be incontestable.

Imagine some photons in a box, Einstein said, and equip the box with a shutter operated by a clock. Let the shutter open for just a moment, at some precisely specified time, so that a single photon escapes. Weigh the box beforehand and weigh it again afterward. From $E = mc^2$, the change in weight gives the energy of the fleeing photon. One version of the Heisenberg principle says that the more accurately you try to measure the energy of some quantum event, the less well you can know the time at which it occurred. In Einstein's new argument, so its author believed, that restriction didn't apply. He could measure the energy of the departed photon, and he knew the time it left the box, and he could make both those measurements independently, as precisely as he wished. He could beat the uncertainty principle, Einstein triumphantly declared.

Léon Rosenfeld, a Belgian physicist who would the following year become Bohr's assistant in Copenhagen, didn't officially participate in the Solvay meeting but came to Brussels anyway to observe the contest. He arrived at the university club, where the participants were staying, just in time to see a beaming Einstein, "followed by a court of lesser fry," returning from the sessions. Einstein sat down and with evident pleasure described his anti-Heisenberg thought experiment "before all those admiring people."

Then Bohr arrived, looking "absolutely like a dog who has received a thrashing, with hanging head." He and Rosenfeld had dinner together, with other physicists dropping by their table. Bohr was "terribly, terribly excited," insisting that Einstein couldn't possibly be right, that it would mean the end of quantum theory. But he was unable straightaway to put his finger on the flaw. Later in the evening he cajoled Einstein in the same way, who serenely paid no attention.

But the next morning it was Bohr who was beaming. Overnight it came to him that Einstein had committed the ironic error of neglecting one of the consequences of his own theory of general relativity. Suppose, Bohr said, the box containing the photons was suspended on some kind of spring balance to gauge its weight. At the moment a photon escaped, he reasoned, the box, reduced in weight, would recoil slightly against gravity. This has two serious implications. First, the slight bouncing of the box produces uncertainty in the measurement of its mass, which translates into uncertainty in the deduced energy of the escaping photon. Second, and more subtly, the motion of the box produces a change in the rate at which its clock runs. This is because, as Einstein had proved a decade and a half earlier, a clock runs at a changing rate as it moves in a gravitational field.

Bohr explained with satisfaction that the product of these two uncertainties, in energy and in time, was precisely what Heisenberg's principle said it should be. Einstein, chagrined to see that in his eagerness to prove Heisenberg wrong he had overlooked his own physics, had no choice but to admit defeat. Bohr did not gloat. In his later account of these events, he cannot bring himself to say plainly that he was right and Einstein was wrong. He emphasizes instead Einstein's repeated acuteness in putting his finger on exactly those points where classical and quantum physics most strikingly take leave of each other. He lauds Einstein's influence in pushing the quantum physicists—he means chiefly himself—to lay bare the characteristics and undoubted oddities of their still-new subject.

Bohr's mannerly praise aside, the fact remains that the crushing blow Einstein aimed at quantum mechanics and the uncertainty principle swung by its target, doing no damage and leaving no mark. Although Heisenberg, Pauli, and the rest had taken only a peripheral role in this intellectual duel, "we were all quite happy and felt that now the game was won," Heisenberg said later.

Defeated in his latest attempt to prove that quantum mechanics was flawed, Einstein reverted to his earlier, more fundamental complaint. Quantum mechanics might be logically coherent—but it could not be the whole truth. Chance, probability, and uncertainty, he insisted, arose from physicists' inadequate understanding of the world they were trying to portray with their theories. The mischievous arguments of Bohr, Heisenberg, and the rest amounted to nothing better than a papering over of difficulties whose true resolution lay elsewhere. One day, he was still convinced, a fuller theory would be found, and quantum mechanics could be consigned to history, along with so many other failed hypotheses.

Members of the Nobel physics committee, struggling to understand whether quantum theory had truly come to stay, made no award in 1931. But then, in a rush of confidence, they presented the 1932 prize to Heisenberg alone and split the 1933 prize between Schrödinger and Dirac. It added to Born's lifelong bitterness that his enunciation of the role of probability in quantum theory did not receive recognition with a Nobel until 1954.

In those same years of the early 1930s, political forces were boiling up that would shortly disperse the founders of quantum mechanics around the world. In early 1933, Adolf Hitler took full power in Germany by manipulating the provisions of the Weimar constitutions and by taking advantage of the complacency of his opponents. Immediately, the Nazis began forcing Jews from the civil service and the universities. Einstein, attacked for years as the icon of Jewish science and the enemy of German culture, and already spending a lot of time on the road, made up his mind to leave Berlin for good. Oxford University wanted to give him a position, as did the California Institute of Technology and the newly founded Institute for Advanced Study at Princeton. Einstein leaned for a time toward California. He had already visited, and thought it a paradise. But like most European intellectuals, he found America fascinating, vigorous, and fundamentally barbaric. He revered, in his own way, the glories of German tradition and culture—not Prussian militarism, certainly not the perverted Aryanness that Hitler ranted about, but the deep and durable German culture of music, philosophy, and science.

Einstein was in California when Hitler took control, and made it clear he would never go back to Germany. He returned to Europe briefly, going only as far as the German embassy in Brussels in order to hand over his German passport and renounce his citizenship. By the autumn of 1933 he was in Princeton, where he would remain until his death. Princeton offered him a quiet haven, freedom from any teaching duties, and the semblance, so the institute's founders hoped, of a refined intellectual center in the best European fashion.

In Germany the press exulted over Einstein's departure. If he had quit the country, it only proved he was not a man Germany wanted. With the most prominent Jew out of the way, the Nazis could start working down the lists. Later in 1933, Born recalled, there came the day "when I found my name in the newspapers among those dismissed for racial reasons." After some globetrotting he ended up in Edinburgh. Pauli, who was ancestrally but not formally Jewish, was by this time safe in Zurich, where he remained until the end of his life. Schrödinger, a professor in Berlin, was not Jewish but found life in Germany increasingly distasteful. He spent a few years in Oxford, then took a position in Graz, Austria, partly so he could return to his homeland, but more pertinently so he could live with his mistress, the wife of another physicist, with whom he had had a daughter in Oxford. His wife, meanwhile, lived in Vienna.

But when the Nazis, in 1938, annexed a far from unwilling Austria, Schrödinger fled again. He took a position at the new Institute for Advanced Studies in Dublin, founded under the guidance of Ireland's mathematically trained prime minister Eamon de Valera.

Many other Jewish physicists fled Germany, or tried to. Their colleagues elsewhere scrambled to find positions for them, not an easy task as anti-Semitism was hardly unknown outside Germany. Added to that, many of those trying to escape had leftish politics. Even Einstein's supporters counseled him to keep his political opinions to himself. He had spoken and written sympathetically of Stalin and the Soviet experiment, and had on occasion made sarcastic reference to American vulgarity and materialism. Plenty of Americans were not eager to see an influx of Jewish communist sympathizers into their country.

In his crazed desire to promote Aryan culture and safeguard Germany from noxious foreign influences, Hitler succeeded in the space of just a few years in destroying Germany's preeminent position in physics. English became the subject's lingua franca. Some German physicists openly celebrated the racial cleansing of their profession, no matter what the immediate cost, while others lamented events without managing to oppose them to any effect. Max Planck, though horrified by the Nazis, believed he could remain in Berlin and use his influence to preserve, as best he could, his country's great scientific heritage.

Early on, before Einstein had officially resigned from the Prussian Academy of Sciences, Planck went to see Hitler, to persuade him that the expulsion of Jews would only harm German science. Hitler raved and threatened but also, Planck managed to believe, promised that Jews were in no real danger. Planck in turn tried to persuade Born and others that they should stay because "in the course of time the splendid things will separate from the hateful." When it became apparent that Einstein was not returning, Planck remonstrated in a letter that his stridency in speaking out against the Nazis was making life harder for those in Berlin who were trying to muddle out some sort of compromise. Einstein, who had always regarded Planck as the soul of probity, found his faith in the decency of the German people knocked down another peg. Planck was only "60 percent noble," he now said.

For Planck, who had lived to regret attaching his signature to the infamous World War I petition in defense of Germany, caution was the only conceivable strategy. He was too old and too patriotic to consider leaving his homeland, but it was becoming impossible to obstruct the Nazis even in small ways. He and Arnold Sommerfeld, the old Prussian who for years had publicly defended Einstein and derided anti-Semitism, were attacked by the leaders of the German science movement as "white Jews," more loathsome in some ways than the real thing, since they chose to support Jewish science despite their lack of a genetic imperative.

Also prominent on the list of white Jews was the name of Werner Heisenberg. He kept quiet about politics, as always, but he strenuously defended relativity and quantum theory, the chief obsessions of those who wished to restore an Aryan version of physics. But his attitude toward Hitler was ambiguous, to put it as kindly as possible. He regarded Hitler as a coarse demagogue leading a gang of uneducated thugs, but at the same time he had more than a tinge of sympathy for the idea that Germany needed strong leadership to restore its pride and strength. After a visit to Germany in the early days of the Hitler regime, Bohr returned to Copenhagen relaying Heisenberg's warmly expressed opinion that things were not going so badly, now that the Führer was taking care of the communists and other unpatriotic extremists.

In any case, how long could Hitler last? During Heisenberg's whole life German governments had come and gone, each one as fragile and fractious as its predecessor. Heisenberg was far from alone among reasonable people—apathetic, incurious people—in thinking that the chaos would blow itself out before too much damage was done.

Heisenberg's disdain for politics had served him well thus far. And the expulsion of Jews created some job opportunities. Göttingen expressed an interest in having him come there, to replace Born. Sommerfeld tried to bring him to Munich. But in both cases the authorities put a stop to the proposed moves. Heisenberg was not one of theirs. He had expressed some carefully worded reservations that good physicists were being forced out of the country, but his private protest to officialdom brought no change in policy, only a formal reprimand. Chastened, he held his tongue. In 1935 he signed an oath declaring his allegiance to Hitler's government, as all civil servants were required to do. He consulted Planck about resigning in protest. That would only mean, Planck told him, that an outright Nazi and a lesser physicist would be appointed in his place. Better for German science, in the long run, to stay on and do what one could.

Which amounted, in the end, to nothing much.

Chapter 15

LIFE-EXPERIENCE AND NOT SCIENTIFIC EXPERIENCE

By the time Hitler succeeded in dispersing German scientific talent around the world, quantum mechanics had already gone global. No country gained more from the exodus of Jewish intellectuals than the United States, but American science was already rising in the world ranks on its own merits. European scientists had been crossing the Atlantic even before 1914, and did so increasingly as international tensions calmed after the war. They frankly admitted that their American adventures brought home gratifying amounts of pocket money, but as the years went by, they could hardly fail to notice the increasing sophistication of the audiences they encountered. Meanwhile, young Americans flocked to Europe to pick up the new physics—one American visitor to Göttingen in 1926 found more than twenty of his fellow countrymen already there—but always with the intention of returning home to build up their own institutions.

Britain's presence in theoretical physics was growing again, although the glories of the nineteenth century were never to be regained. It was American English that supplanted German as the international language of theoretical physics. France, in the person of Louis de Broglie, had made its contribution to quantum mechanics, although French physics in general had declined since the days of Becquerel, Poincaré, and the Curies.

The leading edge of science, in other words, rolled this way and that across national boundaries. It had passed from England to Germany in the early twentieth century, pausing for a while in Munich, Göttingen, reaching out to Copenhagen, then brushing by Cambridge again before moving on to the other Cambridge, and to Chicago, Princeton, and Pasadena. Hitler's stark intrusion had the effect, perhaps, of accelerating what was already a growing continental drift. Schools of science, as of art or music, rarely stay for long in one place.

Nonetheless, it's striking that so much of quantum mechanics arose in Germany during so strange and fraught a phase of that country's history. The Weimar period has acquired in retrospect an exotic tinge, as if an alien sensibility settled for a decade on stolid Germany and then flew away again. This was the Germany of civil discontent and disorder, of brief, crazy art movements, of nightclubs and cabaret, of Bertolt Brecht and Fritz Lang, of plodding socialist realism and the technophile Bauhaus. It was manic and incoherent. Artists flew from one obsession to another, fiercely repudiating the past, even when the past was just six months old. Politics was shaky, art volatile, civil life uncertain and at times desperate. The mother of excess, as Nietzsche put it, is not joy but joylessness.

In physics too this was a time of upheaval. The new rule of

probability overthrew the old order of determinism. Ideas rose and fell in years, sometimes in mere months. Classical physics gave rise to the old quantum theory, which brought forth quantum mechanics, which spawned uncertainty. Inevitably, some sociologically minded analysts have been led to wonder if there was not some greater connection than mere coincidence between the ups and downs of the new physics and the social and intellectual volatility of that period. Did the disorderly, contentious mood of Weimar Germany seep also into scientific thinking and promote the rise of uncertainty?

Scientists routinely deride any such suggestion. Physics, they will say, proceeds for its own reasons. Uncertainty had many roots and antecedents, from kinetic theory to radioactivity to the spectra of glowing bodies. Hard to see any influence of art or politics there. And the scientists who evolved the idea of uncertainty were for the most part politically apathetic and artistically conventional. Heisenberg and Born, on piano and violin, liked to play Beethoven. Einstein preferred Mozart. Bohr didn't care for music at all; he played soccer and tennis, and skied well. Pauli liked to stay out late, but didn't much hang around with artists or musicians. And he was proud of not reading the newspapers.

As much as they might try, though, physicists in Germany in that period could not live in pure monkish isolation from the world around them. They experienced shortages of money and food. They saw violence in the streets. Since university positions were in the hands of civil servants, they must have been at least distantly aware that the government changed from time to time, and tried out different policies affecting research and education. Their thoughts may have been on another plane, but they inhabited a real world.

Even so, it is jolting to find a historian of science, Paul Forman, writing thus: "I am convinced . . . that the movement to dispense with causality in physics, which sprang up so suddenly and blossomed so luxuriantly in Germany after 1918, was primarily an effort by German physicists to adapt the content of their science to the values of their intellectual environment." *Primarily?*

The argument, reduced to a couple of sentences, is this: Germany's collapse in the First World War led to profound disillusion with the past, including not just Bismarckian statecraft and a rigidly structured society but the whole ethos, rooted in science, of determinism and order. There arose in opposition to the old ways a sort of romantic revivalism, embracing nature over the machine, passion over reason, chance over logic. If history, like science, was deterministic, and if that determinism had resulted in Germany's downfall, then evidently some other kind of history was urgently required. Therefore scientists too, to avoid being associated with the discredited past and to curry favor in the new intellectual climate, likewise abandoned determinism and marched under the banner of chance, probability, and uncertainty. According to Forman, "the readiness, the anxiousness of the German physicists to reconstruct the foundations of their science is thus to be construed as a reaction to their negative prestige."

Of course, no physicist would ever admit to proposing a radical new theory in order to conform to some passing social trend. The influence, if such there was, would be subliminal, unconscious, discernible only by a trained and observant historian.

Some scientists, to be sure, had reacted overtly to the changing order brought about by Germany's collapse. Max Planck vocally promoted the cultivation of science as a way for his country to rebuild its lost honor and salvage its international reputation. But Planck was also notable for his lack of enthusiasm about the deeper implications of quantum theory. The power and durability of science, in Planck's view, rested precisely on the robust deterministic foundations established in the nineteenth century, and it was by emphasizing that solidity, Planck believed, that German science could prove its worth. Science, in other words, could exert a benign and calming influence by resisting contemporary pressures and upholding old standards—which is precisely the opposite of saying that science should adjust its principles to gain favor in a volatile world.

Certainly there was, in postwar Germany, a strain of atavistic anti-intellectualism that took aim at the cold, hyperrational scientific view of the world, but like so much else in Weimar Germany this was no coherent philosophy but a welter of impulses. Young men in the Pfadfinder movement, so dear to Heisenberg's heart, tramped around the hills and forests, swooning over the wonders of nature and arguing endlessly about the meaning of life. "Such thinking," the cultural historian Peter Gay has said, "amounted to nothing more than the decision to make adolescence itself into an ideology." In any case, the Pfadfindern were a varied bunch. Some were socialist and wanted to make a new, egalitarian world; some leaned right and yearned for a restoration of the old Germany, where everyone knew their place. Heisenberg and his comrades did not trouble themselves much with contemporary politics, except to bemoan it all. Throughout his early scientific career, when he was formulating novel mathematics and the uncertainty principle, Heisenberg would take off from time to time to wander with his friends among the mountains and lakes. For him, it was pure refreshment, an escape from the tribulations of everyday life. During these outings he only wanted to get away from society, not reform it.

If the inchoate romantic inclinations of this period ever found an intellectual leader, or rather a guru, it was Oswald Spengler, who in 1918 and 1922 published the two volumes of his vast, dense autodidactic work *The Decline of the West* (in German,

Der Untergang des Abendlandes, a far more resonant and doomy title). Spengler was a schoolteacher who toiled every evening to compile his undoubted erudition and learning into a panoramic, all-encompassing theory of world history. He had educated himself, it seemed, about every obscure and ancient culture that had ever inhabited the world's four corners; he had studied and digested their art, their philosophy, their music, their mathematics. His big theme was destiny, or rather the *Destiny-idea*. History followed a great cycle, Spengler said. Cultures rose and fell, and their styles of thinking waxed and waned along with them. The modern scientific, rational culture was just one more turn of the wheel; it too would fall.

Spengler's method is to lay out reams of detail and weighty quantities of obscure facts, and then, as the reader's head begins to nod, to leap adroitly to grand assertions about what it all must mean. It's hard to describe the laboriousness, the solemnity, the tendentiousness, the sheer flat-out daffiness of the Spenglerian enterprise. His gloomy, fatalistic opus became a huge bestseller. For German readers, it offered the consolation that the wheel of history would continue to turn, and a fallen country—and culture—would rise again. It was *Destiny*.

For the present ills of the world, Spengler wrote, science was to blame, going all the way back to the ancient Greeks and their fateful embrace of logic and geometry. Goethe was his hero, Newton the arch villain. Goethe "hated mathematics . . . For him, the world-as-mechanism stood opposed to the world-as-organism, dead nature to living nature, law to form."

Against tiresome, shallow scientific causality stands the historical force of destiny. The former is mere accident, while the latter connotes purpose. "The Destiny-idea," Spengler tells us, "demands life-experience and not scientific experience, the power of seeing and not that of calculating, depth and not intellect . . . In the Destiny-idea, the soul reveals its world-longing, its desire to rise into the light, to accomplish and actualize its vocation."

A little of this goes a long way, and in *The Decline of the West* it goes a long way indeed. To put it simply, Spengler captured the sense that something was dreadfully wrong with the state of the world but that there was a way out, in which the rejection of rationalism, science, and in particular coldhearted determinism would play a large role.

Whether Spengler was truly influential, rather than just popular, is hard to judge. The Nazis adopted his theme of a resurgent culture that rejected modernity, but in an opportunistic way that Spengler himself resented. No scientist could take him seriously. Spengler was not asking for a new kind of science, softer and gentler and less prescriptive than the old stuff. He was against science in all its manifestations.

Forman would have us believe that scientists rejected determinism and causality and embraced uncertainty and probability as a sop toward all those Germans enamored of the kind of thinking Spengler exemplified. But he can provide no real evidence of this, only an assertion that the rise of uncertainty fit in with the tenor of the times. Einstein at least glanced at Spengler, and wrote to Born about the experience: "In the evening one goes along with what he suggests, then smiles about it in the morning . . . Such things are amusing, and if tomorrow someone says the exact opposite fervently enough, that's amusing too, and what's *true*, the devil knows." This is perhaps Einstein's version of Bohr professing to find some novel thought "very interesting."

More fundamentally, uncertainty did not erupt capriciously in the mid-1920s. It had been welling up for a decade or more already by then, forcing itself upon the reluctant consciousness of scientists. When probability and uncertainty took their central roles in quantum mechanics, it was for concrete and specific reasons. These were not whimsical changes to the structure of physical theory, but solutions to deep and difficult problems that had confounded physicists for years.

Nor is it true that quantum mechanics was entirely a German production. Leadership came from Bohr, a thoughtful Dane who admired German science but was not at all enraptured by lofty invocations of German culture and spirit. Crucial contributions came from Dirac in Cambridge, from Kramers, a Dutchman in Copenhagen, from Pauli and Schrödinger, both Viennese, and from de Broglie, a minor Parisian aristocrat.

Nor do the fault lines of politics and characters separating the pioneers and the critics of quantum theory tally neatly with their scientific beliefs. In the anti-probability camp we find Nazi sympathizers such as Johannes Stark, old-fashioned right-wingers such as Willy Wien, and moderate conservatives like Planck, along with the avowedly socialist Einstein and the mostly apolitical Schrödinger. The latter two were arguably the most bohemian of the physicists in their personal lives, and in that respect perhaps the most in tune with the alleged spirit of Weimar, but in physics they led the call for a restoration of the old order. Meanwhile Heisenberg, the originator of uncertainty, was conventional and facile in his politics, rather prim and timid in his personal life — solidly bourgeois, in other words — but in his science he was willing to set aside formal rigor and let his intuition guide him. Pauli was almost the opposite. He was no respecter of reputations and had little use for social nicety, but as he himself admitted, he sometimes let caution and fear of the unknown inhibit his scientific imagination. Not long before he died, Pauli lamented to an interviewer that although he thought he had been a freethinker in those heady days, he realized, looking back, that "I was still a classicist and not a revolutionary."

In short, these are pieces of a jigsaw puzzle that don't fit neatly together. Conceivably, Germany's prominence in the rise of quantum physics had something to do with the emergence of an intensely mathematical kind of theory in that country, as opposed to the more pragmatic school of nineteenth-century British mathematical physicists. But it's not hard to find reasons that seem more arbitrary than predestined. If any single early event was fateful, it was the instant fascination that Sommerfeld showed for Bohr's original system of electron orbits within the atom. Sommerfeld in turn trained Pauli and Heisenberg and a good many others. Dirac, by contrast, hadn't even heard of the Bohr atom until he went to Cambridge almost a decade later. Should we say that it was not the sociopolitical character of the Weimar Republic but rather the idiosyncratic interests of that old hussar Arnold Sommerfeld that made Germany the birthplace of quantum mechanics? But then to what psychological, social, and political factors do we attribute the fact that Sommerfeld was so drawn to the Bohr atom when so many other physicists were perplexed or repelled?

In the rise of uncertainty in Germany, in other words, there's an irreducible element of contingency in addition to discernible intellectual trends. In this respect scientific history is like history in general—unless, as Spengler would have it, it all unfolds from the *Destiny-idea*.

If the forces of unreason didn't compel scientists to introduce uncertainty into physics, it's notable that the idea of uncertainty was quickly embraced by at least one prominent figure who was no friend of science and logic. Only a year or two after Heisenberg came up with his principle, D. H. Lawrence penned this little scrap of poetry:

I like relativity and quantum theories
Because I don't understand them
And they make me feel as if space shifted about like a swan
  that can't settle,
Refusing to sit still and be measured;
And as if the atom were an impulsive thing
Always changing its mind.

Lawrence admired impulsiveness far more than he cared for reason, so it pleased him that the scientists had been hoist, so it seemed, by their own petard. Their effort to understand and predict the world through a complete system of laws and rules had stymied itself. Now they had laws that said they couldn't know everything, that time and space would not conform to their wishes. Spengler the person—a desiccated bachelor hunched over his ancient texts—Lawrence would no doubt have found pathetic, hardly a real man at all, and he would not have had much truck with the elaborately over-theorized system of history Spengler offered. Yet in their dim view of science these opposite types had a connection. Spengler rejected the overweening intellectual determinism of the nineteenth century. Lawrence railed against the coldhearted world of technology and industry (as he had some reason to, having grown up in a grim coal-mining area of England). In different ways, science to them represented something inhuman, debilitating—something that was now overthrown, or at least tottering.

Even the scientists had to agree that complete determinism of the old type was gone. Born said so, and Heisenberg amplified the point. But science, contrary to the desires of the Spenglers and Lawrences of the world, did not abruptly stop working. That was the more interesting puzzle that Bohr in particular applied himself to. His language of complementarity was meant to provide the means by which scientists could continue to speak rationally and consistently of what they did, even though one of the seemingly necessary supports of their subject was showing cracks.

And this was why the question of uncertainty seemed so intriguing and important to so many people outside science. Was science fatally crippled? Would it carry on regardless (which seemed to be the wish of most young physicists, blithely confident that as long as they could calculate, they could do science as they had always done)? Or would it change, and if so how?

Such questions, if fascinating to poets and philosophers, struck no sparks in the great majority of working physicists. But Bohr and Einstein, as always, were the exception. Bohr wanted to show how physics could continue in good conscience, despite the intrusion of uncertainty. Einstein wanted to show it could not. And he had one last trick up his sleeve, one final demonstration that quantum mechanics was not the last word.

Chapter 16

POSSIBILITIES OF UNAMBIGUOUS INTERPRETATION

Before taking up permanent residence in Princeton, Einstein lingered for a few last months in Europe, spending his time mostly in Belgium and England. On June 10, 1933, he delivered a lecture in Oxford on his views of theoretical physics in general and quantum mechanics in particular. The theorist must pay close attention to observational evidence and empirical phenomena, he said, but that was only the first step. In creating theories, the scientist must employ imagination to connect facts into a coherent structure framed according to the rigorous rules of mathematics and logic. That, of course, was how he had arrived so many years before at his theories of special and general relativity.

His guiding principle, Einstein said, was the conviction that nature always chooses the simplest solution. "In a certain sense, therefore," he went on, "I hold it to be true that pure thought is capable of comprehending reality, as the ancients dreamed." But this veers toward a dangerous idea. Einstein, when young, had insisted that the springboard of his imagination must be anchored in scrupulously examined facts. Now, past the age of fifty, Einstein seemed to be saying that intuition and reason alone, divorced from crass practical considerations, could suffice to determine natural laws.

Simplicity in scientific theorizing is often characterized as elegance or beauty. This feeling of aesthetic rightness, whatever it is called, can be a deception as well as a guide. Bohr had his own view on the matter. "I cannot understand," he once said, "what it means to call a theory beautiful if it is not true."

Einstein talked in Oxford of his unease with quantum mechanics—because it didn't comport with what "pure thought" told him of the way a theory of physics ought to work. Born's probability interpretation of quantum waves, Einstein insisted, would have "no more than a transitory significance." He argued that in a more satisfactory theory than quantum mechanics, physical events would regain their traditional objectivity and would not be seen as mere constellations of possibilities. On the other hand, he accepted that the location of a particle, because of Heisenberg's uncertainty principle, could not be given any definite, absolute meaning. Quite how he hoped to reconcile these two contradictory statements he left unsaid.

Once settled in Princeton, Einstein continued to pick away at quantum mechanics. No evidence pointed to practical flaws in the new physics. But his inner voice—or as he liked to put it, the voice of "the Old One," filtering down to him alone—told him that something was seriously amiss. He had listened to this voice before. Why should it fail him now?

In 1935, working with his young Princeton colleagues Boris

Podolsky and Nathan Rosen, Einstein published his last and most famous blast against quantum theory. "Can Quantum-Mechanical Description of Physical Reality Be Considered Complete?" the paper asked in its title. The question was rhetorical. The answer was clearly no, according to Einstein, Podolsky, and Rosen.

The EPR argument is an elaboration of what Einstein had fretted about at the fifth Solvay conference in 1927. There he had seized on Born's assertion that a quantum wave function can describe only the probability of a particle being one place or another. That's all very well, said Einstein, but at some point probability must turn into certainty. An electron hitting a screen, in the example he chose, has to land at one place in particular. And when it lands, must not the quantum wave describing it somehow change instantaneously all across the screen?

No one then had seemed to see what he was getting at. The argument was indeed vague and metaphysical. But Einstein, Podolsky, and Rosen now claimed they had made the objection concrete, turned it into a specific and demonstrable problem. They could pinpoint, so they argued, how quantum mechanics took leave of common sense.

First, in the old, true Einsteinian style, they needed to make absolutely clear what common sense amounted to. Any acceptable theory, they declared, must deal in what they called "elements of physical reality." By this they meant such things as position and momentum, the traditional kinds of quantities the physicist, by time-honored habit, regards as unarguable pieces of information about the physical world.

Very well—but what, actually, constitutes an element of physical reality? It was not an issue scientists had ever spent much time worrying about. So Einstein and his colleagues proposed a formal definition, one that has become famous or notorious, according to one's perspective. If, they said, "without in any way disturbing a system, we can predict with certainty . . . the value of a physical quantity, then there exists an element of physical reality corresponding to this physical quantity."

Think, for example, of the position or momentum of an electron. If you have a way to determine either property without in any way affecting the electron's path or subsequent behavior, then you are entitled to say that the electron's position or momentum is a definite fact, an undeniable datum. An element of physical reality, in other words.

Having set the argument up to their liking, Einstein and his colleagues then proceeded to demonstrate how quantum mechanics runs into trouble. They imagined two particles zooming away in opposite directions from some common origin, with the same speed, so that as soon as you measure the position or momentum of one, you automatically know the position or momentum of the other.

They conceded that an observer making measurements of one of the particles would run afoul of the uncertainty principle. Measure its momentum, and lose knowledge of its position, or vice versa, just as Heisenberg dictates. But now Einstein, Podolsky, and Rosen played their trump card. The whole point of their setup was that any observation of one particle tells you something about the other, and that's where strange things begin to happen.

Measure the first particle's position, and you immediately know the position of the second—even though you haven't looked at it directly. Or measure the first particle's momentum, and you also know the second's—again, without looking at it. Which means, the authors eagerly concluded, that both the position and the momentum of the second particle must be "elements of physical reality." Because these properties can be determined without disturbing the particle in question, they must have definite, preexisting values. It cannot be, they argued, that a measurement on the first particle *only then* causes the second particle's characteristics to materialize out of a quantum fog—because nothing has actually happened to the second particle.

And the larger implication, they went on, is that Heisenberg's vaunted uncertainty principle does not, after all, mean that physical properties are fundamentally indefinite until measured. Rather, particles have definite properties, and the uncertainty principle is an admission that quantum mechanics cannot fully describe those properties. Which means, Einstein and his young collaborators concluded, that quantum mechanics is not telling the whole story—just as Einstein had long insisted. It was a partial theory only, an incomplete portrayal of the underlying physical truth.

In Copenhagen, "this onslaught came down upon us like a bolt from the blue," recalled Bohr's assistant, Léon Rosenfeld; "everything else was abandoned. We had to clear up such a misunderstanding at once." Bohr himself said that the paper's "lucidity and apparent incontestability . . . created a stir among physicists." Schrödinger applauded Einstein for his latest intervention, but everyone else was more irked than fascinated. Pauli wrote to Heisenberg that the EPR paper was "a catastrophe," although he was generous enough to allow that he would regard a young student who had come to him with it as "quite intelligent and promising."

Pauli urged Heisenberg to respond and wondered if he too should "waste pen and ink" trying to set matters straight. In fact, when *The New York Times* ran a story headed "Einstein Attacks Quantum Theory," the reporter found an American physicist who put his finger on the big difficulty. Edward Condon observed that "of course, a great deal of the argument hinges on just what meaning is to be attached to the word 'reality.'"

Heisenberg worked up a reply, but withheld publication when he learned that Bohr was also writing a response, ceding to Bohr his old papal authority to issue pronouncements on questions of dogma. (Bohr, it's not surprising to learn, claimed years later that Heisenberg's proposed response was flawed anyway.)

Naturally, Bohr took his time. With his assistant Rosenfeld he dissected the EPR paper, going back and forth on one proposed rebuttal after another, sometimes coming to a halt in the middle of the tortuous discussions and asking, "What *can* they mean? Do *you* understand it?" Drafted, reworked, and worked over again in his usual fashion, Bohr's agonizingly constructed reply to EPR, published five months later, reveals the Danish master at his awkwardly prolix, exasperating best. The gist of it, Bohr says, is that despite all the metaphysical flourishes, Einstein and his colleagues have not found a practical way to beat the uncertainty principle. Even in the EPR setup, you still can't actually deduce at the same time the position and momentum of either of the particles, directly or indirectly. In any practical sense, Heisenberg's principle stands.

Instead, Bohr explained, the EPR argument begins with a certain definition of physical reality, and then shows that quantum mechanics doesn't stack up. Or in Bohr's words: "The apparent contradiction in fact discloses only an essential inadequacy of the customary viewpoint of natural philosophy for a rational account of physical phenomena of the type with which we are concerned in quantum mechanics." Translated into English, this means that Einstein, Podolsky, and Rosen test quantum mechanics against an inappropriate criterion and, no surprise, find it wanting.

On the other hand, something funny seems to be going on in the EPR experiment, and Bohr was careful not to be very precise about what that funny business might be. He specifically eschewed any implication that measuring the first particle somehow causes the second particle's properties to take on, instantaneously, their appropriate values. Instead, he wrote in a famously opaque phrase, "there is essentially the question of *an influence on the very conditions which define the possible types of predictions regarding the future behavior of the system.*" Which seems to mean, if anything, that the observer's choice of what to measure, not yet acted on, will affect how the particles reveal themselves later.

As to the charge that quantum mechanics is incomplete, Bohr admitted that the observer cannot obtain as much information as a classical physicist would want. But he insisted that quantum mechanics nevertheless offers "a rational utilization of all possibilities of unambiguous interpretation of measurements, compatible with the finite and uncontrollable interaction between the objects and the measuring instruments in the field of quantum theory." Translated again, this means that what quantum mechanics gives is all you're going to get.

When he came to summarize his exchange with Einstein some fifteen years later, for a volume of commemorative essays, Bohr could at least see he might have been clearer. "Rereading these passages," he wrote of his original response to EPR, "I am deeply aware of the inefficiency of expression which must have made it very difficult to appreciate the trend of the argumentation aiming to bring out . . ." *Hard to follow*, another person might write, but Bohr, even when he wanted to say something straightforward, could not help but tiptoe cautiously along, putting off the end of the sentence with as many indirections as he could manage to string together.

It's easier, evidently, to say what's wrong with the EPR argument than to find a clear way to think about it. In a rare plain statement, Bohr said that quantum mechanics demands "a final renunciation of the classical idea of causality." But if classical causality and reality have gone out the window, how are physicists to think instead? To that Bohr had no clear answer, except to recommend his philosophy of complementarity, which in effect meant embracing contradiction rather than trying to resolve it.

Einstein, though, when he responded to Bohr's later summary of their disagreement, could only express his long-standing difficulty with "Bohr's principle of complementarity, the sharp formulation of which, however, I have been unable to achieve despite much effort which I have expended on it." In that, he was with the silent majority of physicists who also found Bohr baffling. Most, though, kept their concerns to themselves. It wasn't that hard, they found, to use quantum mechanics without indulging in philosophical worries about the nature of physical reality.

Unconvinced by Bohr, dismayed by the lack of interest, indeed hostility, emanating from Heisenberg, Pauli, and the rest, Einstein expanded on his concerns about quantum mechanics in letters to Schrödinger, his only sympathetic correspondent. In one letter he imagined the case of a bomb rigged up so as to trigger in response to some unpredictable quantum event. If it was hard enough to grasp what was meant by a quantum state combining the probability that this event will happen along with the probability that it will not, what sense could it possibly make, Einstein asked, to think of a state somehow representing a bomb that's both exploded and unexploded?

In a review published later in 1935, Schrödinger borrowed Einstein's idea but gave it a notorious twist. Einstein's bomb became Schrödinger's cat. This poor creature sat helplessly in an enclosed box, accompanied by a small radioactive sample and a Geiger counter hooked up to a hammer that will smash open a vial of poison. In the course of an hour, Schrödinger stipulated, there's a 50 percent chance that the radioactive sample will trigger the Geiger counter and thereby kill the cat. The radioactive atoms themselves, at that moment, must be described quantum mechanically as being equal parts intact and decayed, since they combine both possibilities. But then, Schrödinger insisted, the cat that's linked to the atom must likewise be described, in quantum language, as equal parts dead cat and live cat. And this must be nonsense, mustn't it?

Even more than the EPR argument, Schrödinger's cat conundrum is either a profound puzzle or a maddening piece of misdirection, according to one's view of quantum mechanics. It was by this time well understood that the Schrödinger wave for an electron in an atom captures the probability of the electron being in one place or another around the nucleus—should one decide to conduct an experiment to look for it. But that is not at all the same, devotees of Copenhagen thinking would insist, as saying that the electron is in some literal sense a little bit here and a little bit there at the same time. Likewise, they would say, Schrödinger's talk of a half-dead, half-alive cat is a misuse of language. The quantum description is an account of what you will see when you open the box and look at the cat—it will be dead or alive, with fifty-fifty probability. That doesn't mean there's literally such a thing as a half-dead, half-alive cat.

The problem, as always, lies in making the translation from a quantum description of possibilities to a classical account of results. Since his lecture at the Como meeting, Bohr had agreed that the observer had a certain amount of freedom in deciding how to make that translation, but he insisted that experience and common sense furnished practical guidance. Meaning, in this case, that it was not illegal to describe a whole cat in quantum terms, but it certainly wasn't very helpful or sensible. Why would anyone want to do that? Bohr's argument, in essence, was that scientists know from experience that measured electrons are in one place or another, and that observed cats are either dead or alive. So what was the problem? What was the point in using inconsistent language to describe the physically impossible state of a cat you haven't looked at?

To Einstein and Schrödinger, of course, it was Bohr who was missing the point. Schrödinger ran into Bohr briefly in London in the spring of 1936 and relayed back to Einstein the news that Bohr, speaking in his careful, charming way, deemed it "appalling" and "high treason" that certain critics should argue so strenuously against quantum mechanics. His objection was specific: Einstein and Schrödinger, Bohr said, were trying to impose their will on quantum mechanics, rather than listening to what quantum mechanics was saying. As Bohr forcefully put it on another occasion: "It is wrong to think that the task of physics is to find out how nature is. Physics concerns what we can say about nature." This is not so far from Wittgenstein's famous closing statement in the *Tractatus Logico-Philosophicus*—"whereof we cannot speak, thereof we must be silent"—although there is no evidence that Bohr ever tackled Wittgenstein's terse, aphoristic volume.

The plaintive cries of Schrödinger's cat, to be fair, called physicists' attention to one crucial issue. How does an uncertain quantum state deliver a definite answer to a classical question? One response to the puzzle was the assertion that human intervention is required: only when an observer looks at the cat is it obliged to become clearly dead or alive. This strangely popular interpretation of quantum events was always a nonstarter. Electrons jumping within atoms and radioactive nuclei decaying are two obvious processes, ruled by quantum uncertainty, that carry on regardless of any observer's attention or lack thereof.

According to Bohr, as always, worrying about such matters is basically pointless. Through long experience, physicists know perfectly well when a measurement has happened. Practically speaking, the cat stays out of the picture. For most physicists, who preferred not to delve too deeply, this was good enough. Heisenberg told Bohr early in the 1930s that "I have given up concerning myself with fundamental questions, which are too difficult for me." And in a lecture series in 1955 at the University of St. Andrews in Scotland, Heisenberg largely endorsed Bohr's recommendations, and said firmly that "we cannot and should not replace these concepts by any others."

Heisenberg's attitude was for many years the norm among physicists. Worrying about metaphysical and interpretational questions arising from quantum mechanics was seen as a low and disreputable occupation. But in 1964, the physicist John Bell came up with an ingeniously simple way to make a feasible, if difficult, experiment out of the EPR argument. Repeated tests on suitably arranged pairs of particles, he showed, would yield a measurable difference between what quantum mechanics ordained and what would follow if EPR's definition of "elements of physical reality" held true. Some two decades later, when these technically demanding tests were made, quantum mechanics proved wholly correct. Einstein's inner feeling for the shape of physical reality had led him down the wrong path.

But this doesn't exactly tidy the dispute away. Bohr's argument was ultimately that to talk of a quantum cat, a strange half-dead, half-alive animal, was just silly. Schrödinger, though, with Einstein's agreement, insisted that nothing in formal quantum theory prevented you from thinking about quantum cats, if you so wished, and unless you understood what was going on here, you couldn't claim to understand how quantum mechanics worked. You couldn't just rule such difficulties out of bounds, as Bohr seemed to want.

On this conundrum recent progress, both theoretical and experimental, has shed some light. A cat, unlike an electron, is not an elementary particle. Its numerous atoms and electrons do not sit quietly in some single quantum state. They bounce around and interact, as nineteenth-century proponents of the kinetic theory of gases knew very well. From a theoretical perspective, to talk of the quantum state of a cat means to specify exactly what every single atom and electron in a cat is doing at some precise moment—and this state changes with unimaginable rapidity from one instant to the next. A cat's quantum state, therefore, is a fickle, elusive thing.

Meanwhile, on the experimental side, laboratory physicists have devised methods that can keep a collection of atoms in a genuine quantum state, fixed and unchanging, but only for a handful of atoms and only for a short period of time. These states, for as long as they can be maintained, exhibit true quantum behavior.

The upshot is that according to modern thinking, Schrödinger's talk of the quantum state of a cat was too glib. If it were possible to maintain all the atoms of an entire cat in a single, fixed quantum state, then it would be possible to speak of half-dead, half-alive quantum cats. But in reality, the endless and unfathomably complex interaction of the cat's atoms is enough to ensure that no such quantum state can exist, except for an uncapturable fleeting instant. Rather, what we observe of a cat can be only those properties that remain fixed while the internal quantum state jounces around this way and that. And those fixed properties, so the argument goes, are precisely what we think of as "classical" cat attributes—its being dead or alive, for instance.

But if Schrödinger was wrong in thinking it made sense to talk of the quantum state of a cat, then so was Bohr wrong in thinking that one could but that it would be absurd. In fact, the quantum state of a cat is a more subtle concept than either man grasped. Still, Bohr was perhaps closer to the truth in his instinctive feeling that real cats do not behave in quantum ways, even though—in typical fashion—he had no very convincing argument why this should be so.

In any case, probability has not disappeared, Schrödinger's cat still has a fifty-fifty chance of being found alive when the box is opened. Beyond that, nothing more can be said. That, ultimately, is what so distressed Einstein—the idea that physical outcomes are truly unpredictable. Physicists today who share that distress cannot shake the feeling that something must be missing, that quantum mechanics must be, as Einstein, Podolsky, and Rosen said, incomplete. On the other hand, no experiment has yet found a flaw in quantum mechanics, and no theorist has come up with a better theory.

Chapter 17

THE NO-MAN'S-LAND BETWEEN LOGIC AND PHYSICS

Philosophy, Paul Dirac once observed, "is just a way of talking about discoveries which have already been made." That captures the hostility of most physicists, who do not take kindly to philosophers telling them what theories mean, still less to those who dare to tell them how to conduct their business. Yet Heisenberg, late in life, offered a remark to the effect that Bohr was at heart more of a philosopher than a physicist. Whether this was meant as criticism or merely observation is hard to tell. Heisenberg himself, once he had gotten over his youthful passion for ontological rambles with his Pfadfinder brethren, evinced little interest in attempts to construct a helpful philosophy of the quantum world.

But Bohr was not like other physicists. Unmathematical, he moved forward on a spiderweb of concepts, principles, and riddles that, to the typical physicist, looked something like philosophy. In his Nobel lecture, Heisenberg paid tribute to his mentor by saying plainly that quantum mechanics arose "from the endeavour to expand Bohr's principle of correspondence to a complete mathematical scheme by refining his assertions." Correspondence—the idea that quantum theory has to match smoothly onto classical physics—was to Heisenberg a broadly philosophical assertion that needed to be cast in quantitative, mathematical form so as to yield a real theory. Likewise, as far as Heisenberg was concerned, Bohr's other great principle of complementarity—the idea that wave and particle behavior are contradictory yet equally necessary—was a largely philosophical notion that on occasions shed light on physical problems. But for Bohr, characteristically, the principles came first. Complementarity in particular became his idée fixe, and Bohr began to see it everywhere, in increasingly grandiose forms.

Almost alone among the pioneers of quantum mechanics, Bohr was willing, indeed eager, to write and speak about the larger meaning of probability and uncertainty, and to speculate on how these changes to the way physicists thought might come to influence other sciences too. (When Einstein wrote and spoke on these broad topics, it was of course with the hope of reining in their pernicious influence, not enlarging it.)

In 1932, Bohr spoke on "Light and Life" at a conference in Copenhagen on the subject of light therapy for various medical problems. A few years later he discussed "Biology and Atomic Physics" at a memorial meeting for Luigi Galvani, the Italian scientist who in the late eighteenth century had made frog muscles twitch by the application of small voltages. By 1938 he was speaking to anthropologists and ethnologists on "Natural Philosophy and Human Cultures." Typically, he would begin by apologizing that he, a mere physicist, presumed to talk about matters beyond his professional expertise. Then he would plow right in anyway.

He introduced his grand idea, complementarity, by explaining briefly how it resolved the conflict between wave and particle depictions of light. Physics now taught that different kinds of observations led to different and even discrepant scientific pictures, and he would urge this principle on his audiences as a lesson for all scientists to consider. In speaking about life, for example, he said that you could think of an organism as an intricately connected collection of molecules, performing their mechanical tasks in accordance with the basic laws of physics, or you could think of the organism as a functioning whole, with attributes we call will and purpose. These were complementary viewpoints, he said, not merely because they offered different perspectives, but because they were impossible to sustain simultaneously. If you want to understand life as an intricate mechanism, he would argue, you have to pick an organism apart molecule by molecule to see how it works, but in doing so, you will lose sight of qualities of life that derive from the organism as a whole. If, on the other hand, you want to study life organically, as a whole, then you cannot hope to tease out the role of every single molecule.

From this observation Bohr jumped to the dramatic assertion that "the concept of purpose, which is foreign to mechanical analysis, finds a certain application in biology." Complementarity, he was saying, meant that purpose could exist as a property of whole organisms, even though it had no meaning in terms of underlying molecular processes and biochemistry. Of course, this rules out of order any questions about where purpose comes from, scientifically speaking, and this kind of evasion is precisely what Einstein found so exasperating when Bohr applied it to questions about the nature of physical reality.

In psychology, Bohr found illumination in complementarity concerning the fact that we are creatures of both reason and emotion. We can analyze with dispassion and logic; at the same time we make choices according to feelings and sentiments that are not rationally explicable. The same brain does both, and although Bohr at that time had no model of brain function that he could link to our reasoning and emotional capacities, he evidently believed that complementarity made it possible for logic and illogic to arise from the same source.

Whether Bohr meant these arguments literally or metaphorically is far from clear, and if pressed, he might well have smilingly responded that meaning and metaphor were complementary aspects of language that must both be kept in mind at all times. According to Rosenfeld, Bohr once said that "whenever you come with a definite statement about anything you are betraying complementarity." It's tempting but alas implausible to think that Bohr might have been making an ironic joke at his own expense.

As Bohr spoke more and more enigmatically on increasingly wide-ranging subjects, his determination not to say anything straightforward or concise begins to seem almost a phobia, a psychological hang-up. Other physicists mostly shook their heads in sad puzzlement. Like any great scientist, Bohr had earned the right to indulge himself a little. So too had Einstein, but at least Einstein tried to stick mostly to specific questions of physics, and to make his objections plain, even if few took him seriously anymore. Bohr was in a world of his own. And though his audiences of biologists, psychologists, anthropologists, and the like no doubt felt honored by the physicist's presence and favored by his deep remarks, there's little evidence that Bohr's views had much influence beyond his own realm of physics.

Whether physicists liked it or not, meanwhile, philosophers of a professional stripe could hardly fail to take note of the strange ideas injected into physics by the quantum pioneers. Uncertainty in physics arrived at a time of considerable uncertainty among philosophers, who were splitting into camps with divergent opinions on what the point of their own studies was. Their attitude toward quantum mechanics in general and Heisenberg's principle in particular likewise split on ideological lines.

Despite being on the losing side in the battle over the reality of atoms, positivist thinking survived and in fact grew more ambitious in the school of thought known as logical positivism, which made its home in the Vienna Circle of the 1920s. The logical positivists proposed to construct a sort of philosophical calculus for science itself. Beginning with empirical facts and data, their system would show how to create rigorously sound theories able to withstand the most stringent philosophical analysis. If science could be made logically foolproof, then its credibility would be beyond question.

Ernst Mach and the older positivists had believed that theories were merely systems of quantitative relationships among measurable phenomena; they did not point the way to some inner truth about nature. The logical positivists, broadly speaking, went along with this idea, but argued that if science couldn't aspire to deep meaning, it could at least hope to attain reliability. And that meant that the language of science must be written in pure, verifiable logic. The positivists' writings of this era are impressively filled with formal equations of symbolic logic and mathematical probability, intended to convince the reader that there is a calculus for concluding that theory A is x percent more trustworthy than theory B in terms of its ability to explain the available data, and further that if some new datum D comes along, then one can turn the wheels of the machinery and test whether D confirms theory A more than it confirms B.

Of course, this bears no relationship at all to what working scientists actually do, but that doesn't appear to be the point. Scientists would continue to invent theories and do experiments in their haphazard, intuitive, heuristic way, and philosophers would act as umpires. But the umpires' rule book turned out not to be as foolproof as its authors had hoped. Carl Hempel, a Vienna Circle member, came up with a tricky difficulty. Suppose your theory is that all ravens are black, Hempel said. Finding a raven of any other color would prove the theory wrong, which is as it should be, and finding a raven that's indeed black lends some degree of support. But a logical oddity arises. The statement that all ravens are black necessarily implies that anything that isn't black can't be a raven. And so, Hempel argued, finding any object that isn't black and isn't a raven—a white elephant, a blue moon, a red herring—amounts to a smidgen of support for the black raven theory. This may be logically inescapable, but it seems an awfully long way from anything resembling science.

Equally seriously, the project of logical positivism, in some sense an exercise in nineteenth-century deterministic thought, got under way just as the physicists were disposing of determinism in their own subject. The uncertainty principle arrived when the philosophical goal of devising a copper-plated scientific method was on its last legs.

Some philosophers, who already believed that the search for an objective account of nature was a delusion, took Heisenberg's principle as evidence that science itself had now confirmed their suspicions. There was no further point, then, in arguing about what scientific theories mean in terms of their relation to some supposed world of facts. The interesting thing instead was to think about how scientists come to agree on their theories, what beliefs and prejudices guide them, how the scientific community subtly enforces the common wisdom, and so on. Such studies have evolved away from philosophy and now go under the name of the sociology of science. One example of this thinking would be Paul Forman's assertion that uncertainty arose as a political response to the conditions of Weimar Germany and had next to no connection with any tedious problems of physics itself.

Among more traditional philosophers, on the other hand, the belief persisted that a rational account of the physical world was not so unreasonable a goal. To such people the uncertainty principle came as unwelcome news indeed. Karl Popper, in his 1934 book *The Logic of Scientific Discovery*, enthusiastically dispatched the ambition of the logical positivism that theories could be proved true, and introduced the now commonplace notion that it is only possible to prove theories false. Theories become more credible, he argued, the more tests they pass, but no matter how well they do, they remain always vulnerable to disproof by some novel experiment. Theories can never gain any guarantee of correctness. Science builds up an increasingly complete picture of nature, but even the most treasured laws of science remain subject to repeal, should the evidence demand it.

Because Popper put so much weight on the ability to test theories, he had to assert that experiments would always yield consistent, objectively reliable answers. Theory might be unreliable in some ineradicable way, but empirical science had to be absolutely trustworthy. And there he ran into trouble with Heisenberg's principle, which said that the sum of all imaginable tests of some quantum system would not necessarily yield a set of consistent results. For his philosophical analysis to work, Popper believed he needed an old-fashioned idea of causality—a certain action always produces, in a wholly predictable way, a certain result. Popper's response to quantum mechanics was simple. Heisenberg must be wrong, he said.

Or rather, that's what he said in the original German edition of *The Logic of Scientific Discovery*. He apologized a little for having the audacity to use philosophical methods to deal with a question in physics, but said that since physicists themselves had been obliged to venture into philosophical territory, he had reason to think that an answer might be found "in the no-man's-land that lies between logic and physics."

Popper made the dubious assertion that quantum mechanics could still be correct even if it were possible to do an experiment that beat the uncertainty principle, and he set out just such an experiment, which he had thought up for himself. This was in the year before the EPR paper appeared. Not until 1959 did his *Logic of Scientific Discovery* appear in English translation, and by that time it included in its appendices a copy of a letter from none other than Einstein, saying that although he too wished to evade the unpleasant implications of quantum mechanics, the experiment that Popper had proposed wouldn't do the job. Even so, Popper added other appendices in which he continued to argue, for a variety of reasons, that Heisenberg's uncertainty principle couldn't possibly be the iron rule that the physicists seemed to think it was.

One of the few contemporary philosophers to take physicists' views seriously was Moritz Schlick, who had taken a doctorate in physics under Max Planck before becoming one of the founders of the Vienna Circle. Schlick corresponded earnestly with Heisenberg to find out what the uncertainty principle really meant, and in 1931 wrote an illuminating essay, "Causality in Contemporary Physics," in which he argued that all was not lost. Dissecting the classical notion of causality, he concluded that it was not a precise logical principle so much as a directive or belief that scientists used as a guide in constructing theories.

The significance of uncertainty, Schlick argued, is that it only *partly* upsets the scientist's ability to make predictions. In quantum mechanics, an event may lead to a variety of distinguishable outcomes, with calculable probabilities for each. Even so, physics still consists of rules about sequences of events—something happens, that sets the stage for something else, then, depending on the outcome, some further possibilities come into play. This is a description based on causal connections, Schlick said, except that the causality has become probabilistic. The fact that things can happen spontaneously doesn't mean that any old thing can happen, at any time. There are still rules.

Schlick's account offers a sort of philosophical compromise congenial in spirit to the Copenhagen spirit promoted by Bohr. The strength of Schlick's analysis was that it provided a loosey-goosey rationale for how physics could continue to work.

For most philosophers, though, loosey-goosey won't do. Those who venture nowadays to write on technical matters of quantum mechanics seem mostly to want to make the deliberately equivocal Copenhagen interpretation go away. They show a remarkable fondness for an alternative interpretation of quantum mechanics worked out in the 1950s by David Bohm, which claims to restore determinism by means of what are called hidden variables. The hidden variables carry additional information about quantum particles, and in examples such as the EPR thought experiment determine in advance what the result of measurements will be. The trouble is, the hidden variables remain just that, hidden. Bohm's system by design conceals the determinism in such a way that no experiment can beat the uncertainty principle or otherwise tease out the extra information that would allow an observer to learn more than standard quantum mechanics permits. Some philosophers profess to find this extremely satisfying, though (as with Bohr and complementarity) they have trouble explaining why. Einstein, among others, was not impressed by the contrived nature of Bohm's reworking of quantum mechanics. "That way seems too cheap to me," he wrote to Max Born.

Over the decades, philosophers, historians, and sociologists have written abundantly on quantum mechanics, and especially on uncertainty, yet the great bulk of this effort misses the mark. Historians and sociologists mostly like to write about the conspiratorial origins of the Copenhagen interpretation, about the way Bohr and his minions forced an incomprehensible idea on a pliant scientific audience. Few philosophers, meanwhile, have followed Schlick's example by trying to take Copenhagen at face value so as to evaluate its merits and difficulties. They seem to find it self-evidently absurd and jump to looking for alternatives.

Meanwhile, physicists in their happy ignorance carry on using and applying quantum mechanics to great effect. Some, to be sure, continue along Einstein's path and insist that a theory of nature that is at heart probabilistic cannot be the last word. But such scientists are not as a rule looking for new ways to interpret the standard version of quantum mechanics; they want to change the theory so as to remedy what they see as its omissions and faults. Philosophical attitudes play little part in these efforts, beyond the elementary thought that physics ought to partake of old-time realism.

As has been true since the 1920s, questions of interpretation and philosophy simply do not arise for the great silent majority of physicists who apply quantum mechanics to their endeavors. In the late nineteenth century, especially among scientists educated in the German tradition, there was a feeling that as theoretical physics advanced, it ought to evolve a philosophy along with it. Nowadays most physicists are reared in the Anglo-Saxon style, steer clear of Plato and Kant, and are belligerently uninterested in what philosophers make of their theories.

Chapter 18

ANARCHY AT LAST

If Bohr's numinous principle of complementarity
failed to conquer physics and made hardly a rip-
ple outside the confines of science, Heisenberg's paradoxically
precise uncertainty principle has ascended to a remarkable level
of intellectual celebrity. In the chaos following the 2003 over-
throw of Saddam Hussein, one ingenious editorialist invoked
Heisenberg by way of explaining why reporters were getting the
big story wrong. Journalists embedded with the troops, he said,
naturally took note of all the problems around them—a broken-
down tank, food and fuel shortages, antagonism with the locals,
miscommunication within the military—and deduced from
these immediate difficulties that the operation as a whole was
foundering. But a version of the uncertainty principle, this commentator said, dictates that "the more precisely the media mea-

sures individual events in a war, the more blurry the warfare appears to the observer." The more you focus on the details, in other words, the less you can see the big picture (this seems closer to complementarity than to uncertainty per se, but never mind).

But really, do we need Heisenberg to help us understand that daily reporting, especially from a combat zone, tends to be piecemeal, incomplete, and inconsistent and that broader themes may get lost in the details? There are at least two hoary old clichés that seem to apply just as well here: journalism is the first rough draft of history, says one, and sometimes you can't see the forest for the trees, goes the other. Nothing quantum mechanical there.

Literary deconstructionists have also made a fetish out of the uncertainty principle. They insist that a text has no absolute or intrinsic meaning but acquires meaning only through the act of being read—and therefore can acquire different meanings depending on who is doing the reading. Just as, in quantum measurements, results come about through an interaction between observer and thing observed, so too, we are invited to think, does the meaning of some piece of literature arise through interaction between reader and text (authors having evidently vanished from this equation).

In a 1976 essay in *The New York Review of Books*, Gore Vidal mocked literary theorists who resort to "formulas, diagrams; the result, no doubt, of teaching in classrooms equipped with blackboards and chalk. Envious of the half-erased theorems—the prestigious *signs*—of the physicists, English teachers now compete by chalking up theorems and theories of their own." In particular he talked of how critics of a certain intellectual stripe like to claim Heisenberg's "famous and culturally deranging principle" as justification for their axioms. It is as if literary critics were belatedly trying to accomplish a version of what the logical positivists failed to do half a century earlier. The positivists wanted to make the philosophy of science itself scientific. The critics want to turn the presumably aesthetic business of judging novels into a formally analytical exercise.

Vidal's reference to the uncertainty principle as "culturally deranging" drew a response from a reader knowledgeable about physics, who protested that Heisenberg's statement was a scientific theorem about making certain kinds of measurements and that any application beyond those prescribed bounds was foolish. But Vidal was right. Whether physicists like it or not, Heisenberg's principle has spread far and wide and caused cultural derangement. This has nothing to do with whether quantum mechanical uncertainty has some genuine meaning in various far-flung regions of intellectual study. It has to do with the way Heisenberg has become a touchstone, a badge of authority, for a certain class of ideas and speculations.

The television series *The West Wing* offers a dramatic re-creation of the fast-talking, quick-thinking operatives who inhabit the highest levels of the Washington political scene. In one episode, these fictional characters are being trailed by an even more fictional (metafictional?) camera crew filming material for a documentary about life in the White House. This was a satisfyingly postmodern exercise: a real film crew recording the activities of a fake film crew taping the action of fictional characters in order to make what is, in the fictional world, a *real* nonfiction movie.

At one point in the story, the unseen filmmaker is waiting with C. J. Cregg, White House press secretary, to see if they can nose in on a high-level meeting including the president and the head of the FBI. The filmmaker asks C.J. if this has been a typical day thus far.

"Yes and no," C.J. replies.

"Because we're here?"

"I don't have to tell you about the Heisenberg principle."

"The act of observing a phenomenon changes it?"

"Yes," C.J. says, and they hustle into the meeting.

Throughout the episode, characters are constantly whispering to each other, sneaking away from the cameras, huddling together in quiet corners—all to avoid the perturbing influence of the would-be documentarians. It's hard to conduct political intrigue when people are watching. But this is easy to understand. Put a bunch of cameras in the middle of a tense and private situation, and people will start acting oddly. No one who has taken photographs at a wedding or tried making a home movie of a family reunion will be surprised by this. Why drag Heisenberg into it?

The common element in these examples is the notion that there's no such thing as absolute truth, that what you see varies according to what you are looking for, that the story depends on who is listening and watching as well as who is acting and talking. There's at least a metaphorical connection with what Heisenberg said about conducting measurements. In this sense, if we must blame anyone for the curse of relativism that supposedly afflicts modern thought (no one's story is "privileged," as the sociologists like to say, above anyone else's; all viewpoints are equally valid), then probably we should blame Heisenberg more than Einstein. Relativity—the scientific theory of space-time, that is—indeed says that different observers will see events in different ways, but it also offers a framework by which these different viewpoints can be reconciled to a consistent and objective account. Relativity doesn't deny that there are absolute facts; that's what the uncertainty principle does.

But even in physics, the uncertainty principle is by no means of ever-present relevance. The whole point of Bohr's program of complementarity was to help physicists handle the evident fact that the real world, the world of observations and phenomena in which we live, seems to be pretty solid *despite* the fact that underneath it all lies the peculiar indeterminacy of quantum mechanics. If Heisenberg's principle doesn't enter all that often into the thinking of the average physicist, how can it be important for journalism, or critical literary theory, or the writing of television screenplays?

We already *know* that people act awkwardly in front of cameras, that they don't tell their stories to a newspaper reporter the same way they would tell them to a friend. We *know* that an anthropologist dropping in on some remote village culture becomes the focus of attention and has trouble seeing people behave as they normally would. We *know* that a poem or a novel or a piece of music doesn't mean the same thing to all readers and listeners.

The invocation of Heisenberg's name doesn't make these commonplace ideas any easier to understand, for the simple reason that they're perfectly easy to understand in the first place. What fascinates, evidently, is the semblance of a connection, an underlying commonality, between scientific and other forms of knowledge. We return, in this roundabout way, to D. H. Lawrence's jibe about relativity and quantum theory—that he liked them precisely because they apparently blunted the hard edge of scientific objectivity and truth. We don't have to be as intellectually philistine as Lawrence to see the attraction here. Perhaps the scientific way of knowing, in the post-Heisenberg world, is not as forbidding as it once seemed.

It was the classical dream of perfect scientific knowledge, of strict determinism and absolute causality, that caused alarm when extrapolated beyond the borders of science. Laplace's ideal of perfect predictability—that if you knew the present exactly, you could predict the future completely—made humans, it would seem, into helpless automatons. Think of Marx and Engels and scientific socialism, the assertion that human history unfolds according to inexorable laws. Think of the eugenics movement and its calculated pronouncements about how human beings can be improved through forcible rather than natural selection. The rebellion of thinkers as diverse as Oswald Spengler and D. H. Lawrence against the technocratic dream may not always have been well reasoned, but it came from a powerful and by no means unreasonable fear of scientific overreaching.

But even at its acme, scientific determinism, as we have seen, was never as all-conquering at it seemed. Statistical reasoning, introduced into physics long before Heisenberg was born, made perfect predictability unattainable. On that point our astute observer Henry Adams began to worry that the newly minted power of science, which he saw as both impressive and fearsome, might crumble to nothing. "He found himself," the author writes toward the end of *The Education of Henry Adams*, "in a land where no one had ever penetrated before; where order was an accidental relation obnoxious to nature; artificial compulsion imposed on motion; against which every free energy of the universe revolted; and which, being merely occasional, resolved itself back into anarchy at last."

Amid this intellectual conflict, the emergence of the uncertainty principle of quantum mechanics, a couple of decades after Adams closed his memoir, offered a measure of reassurance to both sides. It set a tombstone on strict classical determinism. At the same time it failed to undermine science in any far-reaching sense. It suggested that science, for all its marvelous power and scope, had limits. Cold rationality would not, after all, supplant all other forms of knowledge.

Therein lies the metaphorical appeal of Heisenberg's uncertainty principle. It doesn't make journalism or anthropology or literary criticism scientific. Rather, it tells us that scientific knowledge, like our general, informal understanding of the everyday world we inhabit, can be both rational and accidental, purposeful and contingent. Scientific truth is powerful, but not all-powerful.

Adams's fear of anarchy was overstated. Pragmatically, physicists carry on doing physics without feeling any great metaphysical unease over the contamination of their subject by probability and uncertainty. They mostly steer clear of deep questions about the meaning of quantum mechanics. As John Bell and his colleague Michael Nauenberg nicely put it on one occasion, "the typical physicist feels that [such questions] have long been answered, and that he will fully understand just how if ever he can spare twenty minutes to think about it."

Bohr's recommendation was not to think too much about it in the first place. He insisted that it makes no sense to ask what the quantum world really looks like, because any such attempt inevitably means trying to describe the quantum world in familiar, that is to say classical, terms, which merely restates the original question. Expressing quantum truths in classical language is necessarily a compromised endeavor but, according to Bohr, it's the best we can do.

One doesn't have to be Einstein to find this not just unsatisfactory but antithetical to the true spirit of science. Where does it say there are questions not to be asked, subjects not to be broached?

Indeed, the progress of science over the last couple of centuries has seen its relentless expansion into areas previously thought off-limits to natural philosophers. Before the latter half of the nineteenth century, questions about the origin of the sun and the earth were the province of theologians. But then scientists, armed with their new knowledge of energy and thermodynamics, smartly annexed this territory. Nowadays, physicists write dense, difficult papers on the origin of the universe itself. In dealing with that cardinal event, these physicists have to grapple with gravity, particle physics, and quantum mechanics all at the same time—except that they have, so far, no unifying theory with which to tackle the difficulties they encounter. Gravity, in the form of general relativity, remains essentially classical in form, assuming smoothness, continuity, and causality in space and time, down to infinitesimally small scales. Quantum mechanics proceeds from discreteness and discontinuity to uncertainty, and in the big bang those two ways of thinking collide.

Physicists have as yet no quantum theory of gravity to guide them as they try to reconstruct how the universe began. Nonetheless, it seems inescapable that the birth of the universe was a quantum event, so that our very existence ultimately hinges on the awkward question of how elusive quantum transformations can generate phenomena that we see as solid and tangible.

If Bohr's position is that such questions can never be satisfactorily formulated, let alone answered, then he seems to be saying that inquiring into the birth of the cosmos is beyond the scope of science. This, to physicists today, simply won't do.

High-end journals of theoretical physics today are filled with attempts to marry quantum mechanics and gravity. The proposals have involved arcane theories based on supergravity, superstrings, extra space-time dimensions, and much else besides. Nowadays the talk is of M-theory and branes, fearsome mathematical structures that few understand, whose existence is not entirely assured, and which in any case have yet to show that they can do the job required of them.

Such efforts have for the most part focused on the microscopic aspect of the problem. That is, physicists want a theory that describes the gravitational interaction between two elementary particles in a quantum mechanical way. But general relativity is not just a theory of gravity. It is also a theory about space and time and causality. It includes the stipulation, to Einstein a bedrock principle, that gravitational influences, like all other physical effects, can go from one place to another no faster than the speed of light.

That's why Einstein fixed on EPR-type experiments as a deep indication that quantum mechanics couldn't be right—because in such situations it seems that some elusive but instantaneous influence connects the quantum behavior of two particles no matter how far away from each other they fly. This uncomfortable long-distance connection, like so much else that's strange about quantum mechanics, arises because of the inescapability of uncertainty. Because the outcome of a measurement on one particle cannot be completely predicted, the second particle has to remain linked in some way, so it seems, in order that measurements on it remain coherent with observations of the first.

So uncertainty upsets the old order not just on the smallest scales, in the way we can find out about individual elementary particles, but also on the cosmic scale, in terms of the way causality and probability connect up across vast distances. A true quantum theory of gravity would—presumably—make sense of all these difficulties.

But it hardly seems likely, at this stage of the game, that in a quantum theory of gravity uncertainty would fade away. All the evidence suggests that it's here for the duration. There can be no going back to the old days of absolute determinism, when, as the Marquis de Laplace hoped, knowledge of the present would bring complete knowledge of the past and the future.

Cosmically speaking, that may be a good thing. The Laplacian universe can have no moment of birth, because any set of physical conditions must arise, logically and inevitably, from some prior situation, and so on ad infinitum. Nothing uncaused can happen.

But the quantum universe is different. Ever since Marie Curie wondered at the spontaneity of radioactive decay, ever since Rutherford asked Bohr what made an electron jump from one place in an atom to another, the recognition has grown that quantum events happen, ultimately, for no reason at all.

So we reach an impasse. Classical physics cannot say why the universe happened, because nothing can happen except that prior events caused it to happen. Quantum physics cannot say why the universe happened, except to say that it just did, spontaneously, as a matter of probability rather than certainty. Einstein was right, in other words, when he complained that quantum mechanics could offer only an incomplete picture of the physical world. But perhaps Bohr was even more right in his belief that this incompleteness was not just unavoidable but actually necessary. We come to a paradox that Bohr would have loved: it's only through an initial, inexplicable act of quantum mechanical uncertainty that our universe came into being, setting off a chain of events that led to our appearance on the scene, wondering what original impetus led to our existence.

POSTSCRIPT

In 1954, the year before he died, Einstein was visited in Princeton by Heisenberg, just for a few hours. The old man was clearly sinking. He was seventy-five, and had known for some years that an abdominal aneurysm was swelling slowly within him. Surgery would have been risky, and Einstein saw no point in trying to stave off the inevitable. He had suffered through a bout of anemia but recovered. When Heisenberg came by, they spoke politely of small matters. Not about the war, and not much about quantum mechanics. "I don't like your kind of physics," Einstein told his visitor. "There's consistency, but I don't like it."

The war had further strained an already distant relationship. Einstein, of course, signed the famous letter to President Roosevelt outlining the possibility of an atomic bomb, but took no part in designing or building it. Bohr had stayed in German-occupied Copenhagen as long as he could before being spirited away, almost fatally, by the Royal Air Force. Although he had written about the physics of fission, Bohr played only an indirect role in the Manhattan Project.

Heisenberg, meanwhile, remained in Germany. His disastrous visit to Bohr in 1941, which broke any remaining friendship between the two men, is the pivot of Michael Frayn's sharp and melancholy play *Copenhagen*. There was some sort of German project to make use of nuclear power; Heisenberg was involved; he sounded out Bohr—perhaps—on some aspects of the relevant physics.

Bohr's wife said that there was always an aloofness, a distance, to Heisenberg's relationships. Her husband had had awkward moments with Heisenberg, she said, but "in between he was a pleasant man . . . He was what you call well-bred. I mean he had nice manners and was pleasant in that way. But there were difficulties with Heisenberg." He had always been a shy, reserved, formal man and never warmed up fully to others. Dirac, hardly the most sociable type himself, found Bohr easy to get on with, thought the waspish Pauli positively amiable, but remained a little uncomfortable around Heisenberg.

What Germany's wartime nuclear program accomplished, or what it tried to, has never become entirely clear. The country was depleted of resources, including intellectual resources, as so many of the physicists nurtured there had been hounded out. Heisenberg, undoubtedly one of the great innovators and conceptualizers in theoretical physics, was not the man to do practical nuclear physics or engineering. It appears he never figured out correctly how a bomb would work and thought a ton of uranium would be needed. Later, in ugly fashion, this failure transmuted into a story that the Germans, meaning in particular

Heisenberg, had turned away from the moral repugnance of building atomic weapons, or had even deliberately misled their political superiors about the feasibility of doing so. Heisenberg never exactly said this. He never exactly denied it.

Many physicists shunned Heisenberg after the war. Bohr tried to be at least cordial. Slowly, Heisenberg worked his way back into the scientific community, eventually becoming director of the Max Planck Institute in Munich. Einstein was long gone by then. Pauli died suddenly in 1958, Bohr in 1962. Heisenberg died in Munich in 1976.

ACKNOWLEDGMENTS

I owe a great debt to the numerous authors who over the years have sifted through the history of quantum mechanics in far more detail than I was able to, and whose work I have relied on enormously in writing my own account. I particularly acknowledge the efforts of Abraham Pais and David Cassidy. Of course they and others are not to be held responsible for any errors or idiosyncrasies in my version of history.

I could not have researched this book without the use of the Niels Bohr Library of the Center for History of Physics, at the American Institute of Physics in College Park, Maryland. Many thanks to the always helpful staff there. I am also grateful to have had easy access and helpful assistance from the Library of Congress, the libraries of the University of Maryland and George Mason University, and the Smithsonian's Dibner Library of the History of Science and Technology at the National Museum of American History. (And thanks to Mary Jo Lazun for putting me on to that last one.)

I had an enjoyable and illuminating conversation about the EPR paper with Abner Shimony of Boston University. Ralph Cahn helped me out with some of the translation from German.

My agent, Susan Rabiner, encouraged (I might even say pushed) me to work out more clearly what this book was about before I started writing it, and without her help, as always, the project would never have got off the ground. The acute editorial eye of Charlie Conrad at Doubleday made the book slimmer, sharper, and more purposeful than it would otherwise have been. Many thanks to both.

For moral support, especially in the uncertain early stages of putting this project together, I thank Peggy Dillon.

NOTES

In these notes I have not attempted to annotate every scrap of information in the text. Details of the participants' lives and works come generally from the works cited in the bibliography, the *Dictionary of Scientific Biography* edited by C. C. Gillispie being the default reference for lesser characters.

For my understanding of the emergence of quantum theory, I relied heavily on the three books by Abraham Pais cited in the bibliography. Cassidy's biography of Heisenberg and Dresden's of Kramers were also useful, as was the lengthy introduction by van der Waerden to his compilation of important papers. The multivolume history by Mehra and Rechenberg I made less use of, only because it goes into far more technical detail than I needed for my telling of the story.

The AHQP interviews are the invaluable oral histories recorded as part of the Archives for the History of Quantum Physics, a joint project, begun in 1960, of the American Philosophical Society and the American Physical Society. (See www.amphilsoc.org/library/guides/ahqp for more details.) I consulted transcriptions of these interviews at the Niels Bohr Library of the Center for History of Physics, at the American Institute of Physics in College Park, Maryland. Most of the AHQP interviews, even of non-native English speakers, were conducted in English, whence the occasional awkwardness of some of the phrasing.

Wherever possible I tried to find the original German sources for remarks quoted in the text, and my translations are therefore sometimes a little different from versions published elsewhere in English.

Bohr, *CW*, refers to Bohr, *Collected Works*.

1: Irritable Particles

p. 10: "a walking catalogue": Remark is by Edward Parry, a future arctic explorer, quoted by Patrick O'Brian in *Joseph Banks: A Life* (Chicago: University of Chicago Press, 1987), 300.

p. 10: *Charles Darwin, before he was married:* N. Barlow, ed., *The Autobiography of Charles Darwin* (London: Collins, 1958), 103–4.

p. 10: *In June 1827, Brown began a study:* I have mixed here Brown's words and observations from his two famous papers in the *Philosophical Magazine* 4 (1828): 161 and 6 (1829): 161.

p. 11: *"The motion of most of these animalcules":* From a letter from Leeuwenhoek to Henry Oldenburg, secretary of the Royal Society, Sept. 7, 1674, in C. Dobell, ed., *Antony van Leeuwenhoek and His "Little Animals"* (New York: Dover, 1960), 111.

p. 16: *"Brown's new thing":* George Eliot, *Middlemarch*, ch. 17; Nelson 2001, 9, was my source for this reference.

p. 17: *It's strangely difficult:* See J. Delsaulx, *Monthly Microscopical Journal* 18 (1877): 1; and J. Thirion, *Revue des Questions Scientifiques* 7 (1880): 43.

p. 18: "une trépidation constante et caractéristique": L.-G. Gouy, *Comptes Rendus* 109 (1889): 102.

2: Entropy Strives Toward a Maximum

p. 20: *"this phenomenon seems to have barely attracted"*: L.-G. Gouy, *Comptes Rendus* 109 (1889): 102.

p. 22: *"We may regard the present state"*: Laplace's famous statement, from his 1812 *Théorie Analytique des Probabilités*, can be found in J. H. Weaver, ed., *The World of Physics* (New York: Simon & Schuster, 1987), vol.1, 582.

p. 26: *"The observed motions of very small particles"*: Lindley, 212; the remark is from Boltzmann's reply to criticism by E. Zermelo.

p. 28: *"it is possible that the motions to be discussed here"*: A. Einstein, *Annalen der Physik* 17 (1905): 549.

p. 30: *"the scientific synthesis commonly called Unity"*: Adams, 431.

3: An Enigma, a Subject of Profound Astonishment

p. 37: *"but Radium denied its God"*: Adams, 381.

p. 38: *"Radioactivity is an atomic property"*: Pais 1986, 55, quoting an 1898 paper by the Curies and G. Bémont.

p. 38: *"The spontaneity of the radiation is an enigma"*: Quinn, 159, quoting the Curies' report to the International Congress, Paris, 1900.

p. 40: *"I have never had a student with more enthusiasm"*: From the Rutherford collection at Cambridge University Library, MS.Add. 7653:PA.296.

p. 41: *What Rutherford and Soddy proposed*: E. Rutherford and F. Soddy, *Philosophical Magazine* 4 (1902): 370 and 569.

p. 43: *The atom might have internal components*: A. Debierne, *Annales de Physique* 4 (1915): 323; a similar proposal is by F. A. Lindemann, *Philosophical Magazine* 30 (1915): 560.

4: How Does an Electron Decide?

p. 45: *Thinking hard, his features slack*: J. Franck AHQP interview.

p. 45: *Bohr had difficulty with English manners*: Sources for Bohr's time at Cambridge are the AHQP interviews with Niels and Margrethe Bohr; Bohr's letters in Bohr, *CW*, vol. 1; and Pais 1986, 194–95.

p. 46: *"quite the most incredible event in my life"*: E. de Andrade, *Rutherford and the Nature of the Atom* (New York: Doubleday, 1964), 11. These much-cited words are said to be from a lecture by Rutherford, but no further details are given. Eve, 197, has Rutherford making a comparison to a rifle bullet bouncing off a sheet of paper.

p. 50: *the idea of energy quanta*: Bohr AHQP interview.

p. 52: *"no attempt at a mechanical foundation"*: Bohr, *CW*, vol. 2, 136.

p. 52: *"Yes, I have looked at it"*: Lord Rayleigh's remark to his son, R. J. Strutt, given by Strutt, *Life of John William Strutt, Third Baron Rayleigh* (Madison: University of Wisconsin Press, 1968), 357.

p. 53: *"There appears to me one grave difficulty"*: Rutherford to Bohr, March 20, 1913, Bohr, *CW*, vol. 2, 583.

p. 54: *"The statistical law is nothing but the Rutherford law of radioactive decay"*: Pais 1991, 191, quoting a 1916 paper by Einstein. The simple thought experiment analyzed here by Einstein was remarkably productive. In this paper he also proved that as well as spontaneous emission of light by atoms in excited states, there must be a process of so-called stimulated emission, in which the probability for an atom to emit a quantum of light is enhanced by the external presence of radiation of the same frequency. This observation, half a century later, became the theoretical basis for masers and lasers.

p. 54: *"That business about causality"*: Einstein to Born, Jan. 27, 1920, Born, Born, and Einstein, *Briefwechsel*.

5: An Audacity Unheard Of in Earlier Times

p. 56: *"randomly chosen numbers"*: Harald to Niels Bohr, autumn 1913, Bohr, *CW*, vol. 1, 567.

p. 56: *"all nonsense . . . just a cheap excuse"*: This and Born's remark are from Landé's AHQP interview.

p. 56: *"unquestionably a great achievement"*: Sommerfeld to Bohr, Oct. 4, 1913, Bohr, CW, vol. 2, 603.

p. 57: *"It is the custom in Germany"*: Pais 1991, 165, from a 1961 interview not in the AHQP collection.

p. 59: *"Bohr confirmed to me"*: Heisenberg 1989, 40.

p. 60: *"I do not believe I have ever read anything with more joy"*: Bohr to Sommerfeld, March 19, 1916, Bohr, CW, vol. 2, 603.

p. 61: *"almost like a second father"*: Rutherford Memorial Lecture 1958, Bohr, CW, vol. 10, 415.

p. 61: *"I am at present myself most optimistic"*: Bohr to Rutherford, Dec. 27, 1917, Bohr, CW, vol. 3, 682.

p. 61: *"play games with their symbols"*: Eve, 304.

p. 64: *"as long as German science can continue in the old way"*: Heilbron, 88.

p. 67: *"now work with an audacity unheard of in earlier times"*: Pais 1991, 88, quoting a 1910 paper by Planck.

p. 67: *Robert A. Millikan carefully measured the photoelectric effect*: Millikan, *Physical Review* 8 (1916): 355; quotations are from pp. 388 and 383, respectively.

6: Lack of Knowledge Is No Guarantee of Success

p. 70: *"a stronger personality than was the Catholic priest"*: Von Meyenn and Shucking; remark is from a letter from Pauli to Carl Jung, March 31, 1953.

p. 70: *Pauli wanted to hew strictly to the experimental data*: Heisenberg AHQP interview.

p. 70: *"conscience of physics"*: This was a widely known sobriquet for Pauli, mentioned by Enz and many others. I haven't been able to pin down who said it first.

p. 71: *"Munich was in a state of utter confusion"*: Heisenberg 1971, 8.

p. 72: *"a downright amazing specimen"*: Sommerfeld to J. von Gietler, Jan. 14, 1919; quoted by Enz, 49.

p. 73: *"What we are listening to nowadays"*: Sommerfeld's preface to the first edition of his *Atombau und Spektrallinien* (Braunschweig: F. Vieweg and Sohn, 1919).

p. 74: *Pauli referred to Sommerfeld . . . as a hussar colonel*: Heisenberg 1971, 24, and AHQP interview.

p. 75: *"A kind of market place to exchange views"*: Heisenberg AHQP interview.

p. 76: *"it's much easier to find one's way"*: Heisenberg 1971, 26.

p. 76: *"My original wish to study mathematics"*: Heisenberg 1989, 108.

p. 77: *"if someone were to say that I had not been a Christian"*: Cassidy 1992, 13.

p. 77: *"when our families had long since eaten their last piece of bread"*: For Heisenberg's youthful life in Munich, see Heisenberg 1971, ch. 2, and AHQP interview.

p. 78: *"the cocoon in which home and school protect the young"*: Heisenberg 1971, 1.

p. 78: *"such a temporary style of life"*: From ch. 14 of *Doctor Faustus*, in the recent translation by John E. Woods (New York: Vintage International, 1999).

p. 79: *"In that case you are completely lost to mathematics"*: Heisenberg 1971, 16.

p. 80: *"I have grasped the theory with my brain"*: Heisenberg 1971, 29.

p. 81: *As Landé put it many years later*: Landé AHQP interview.

p. 82: *"It works fine, but the foundation of it is quite unclear"*: Sommerfeld to Einstein, Jan. 11, 1922, Einstein and Sommerfeld, *Briefwechsel*.

p. 82: *"damn it, I can see that it's right"*: Heisenberg AHQP interview.

7: How Can One Be Happy?

p. 84: *Einstein and Bohr subsequently exchanged little mash notes*: Einstein to Bohr, May 2, 1920; Bohr to Einstein, June 20, 1920; Bohr, CW, vol. 3, 634.

p. 85: *"each one of his sentences revealed a long chain of underlying thoughts" and following remarks:* Heisenberg 1971, 38–39.

p. 87: *the correspondence principle "cannot be expressed in exact quantitative laws"; Pais . . . comments mysteriously that "it takes artistry to make practical use of the correspondence principle." Segrè explains that it amounted to saying, "Bohr would have proceeded in this way":* Pais 1986, 247, quoting a book by H. A. Kramers and H. Holst and following with his own remark; Segrè, 125.

p. 88: *"generalities and matters of taste . . . very interesting":* Cassidy 1992, 130; from a letter from Heisenberg to his parents.

p. 89: *"young Pauli is very stimulating":* Born to Einstein, Nov. 29, 1921, Born, Born, and Einstein, *Briefwechsel.*

p. 90: *"I was, from the beginning, quite crushed by him":* Born AHQP interview.

p. 90: *"I was so impressed by the greatness of his conception":* Born 1968, 30.

p. 90: *"quite different; he was like a little peasant boy when he came":* Born AHQP interview.

p. 91: *"I always thought mathematics was cleverer than we are":* Ibid.

p. 91: *"Born was very conservative in some ways":* Heisenberg AHQP interview.

p. 92: *"How can one be happy":* Pauli, *Science* 103 (1946): 213.

p. 93: *"some of us had begun to feel":* Heisenberg 1971, 35.

8: I Would Rather Be a Cobbler

p. 94: *"Dr. Nils Bohr" and subsequent comment: New York Times,* Nov. 7 and 16, 1923.

p. 96: *"This remarkable agreement between our formulas and the experiments":* A. H. Compton, *Physical Review* 21 (1923): 483.

p. 96: *no one in Germany read the* Physical Review: Heisenberg AHQP interview.

p. 97: *"Bohr is Allah and Kramers is his prophet"*: A widely reported remark; see Enz, 36.

p. 98: *a story unearthed only recently*: Dresden, 292.

p. 100: *Slater said how thrilled he was*: Pais 1991, 235.

p. 101: *"We will assume that a given atom"*: From the BKS paper, included in van der Waerden.

p. 101: *"utters his opinions like one perpetually groping"*: Pais 1991, epigraph.

p. 102: *"You could never pin Bohr down"*: Rosenfeld AHQP interview.

p. 102: *"Quite artificial"*: Pauli to Bohr, Oct. 2, 1924, Bohr, *CW*, vol. 5, 418.

p. 102: *"I would rather be a cobbler"*: Einstein to Born, April 29, 1924, Born, Born, and Einstein, *Briefwechsel*.

p. 102: *"Can you explain to me what the BKS theory was?"*: Born AHQP interview.

p. 104: *"It seems to me the most important question is this"*: Pauli to Bohr, Feb. 21, 1924, Pauli, *Briefwechsel*.

9: Something Has Happened

p. 106: *At Heisenberg's oral exam in July*: Heisenberg AHQP interview.

p. 107: *Born . . . published a paper calling for a new system of "quantum mechanics"*: The paper, titled "Quantum Mechanics," is included in van der Waerden.

p. 108: *Rutherford, no lazybones, commended Bohr*: Pais 1991, 261, quoting a letter from Rutherford to Bohr, July 18, 1924.

p. 108: *"right now physics is very confused once again"*: Pauli to R. Kronig, May 21, 1925, in Pauli, *Briefwechsel*. Ironically, this lament from Pauli came at about the time of his most memorable achievement in physics. Thinking further on the half-quantum number that Landé and Heisenberg had devised for certain atomic transitions, Pauli concluded that it must correspond to a certain *Zweideutigkeit* (ambiguity or two-valuedness) in the electron itself. In effect, he proposed a fourth quantum number, intrinsic to the electron rather than a property of electron orbits, which could take one of two values. Pauli was then led to his famous exclusion principle, which states that each electron in an atom is specified by a unique combination of the four quantum numbers, so that no two electrons can occupy the same state. A little later, Pauli's *Zweideutigkeit* was interpreted by S. Goudsmit and G. Uhlenbeck as the spin of the electron, which comes in half values when compared to the orbital angular momentum any electron can possess. In this tortuous way, it turns out that Heisenberg's half quantum was not so far off the mark after all.

p. 108: *"Now everything is in Heisenberg's hands"*: F. C. Hoyt AHQP interview.

p. 109: *"was always a perfect gentleman"*: Heisenberg AHQP interview.

p. 109: *"completely shocked . . . I got quite furious"*: Ibid.

p. 109: *"Things always go very oddly with him"*: Pauli to Bohr, Feb. 11, 1924, Pauli, *Briefwechsel*.

p. 110: *"The idea suggested itself"*: Heisenberg 1958, 39.

p. 114: *"Well, something has happened"*: This and other details of Heisenberg's stay on Helgoland come mainly from his AHQP interview.

p. 114: *he had written what he called a "crazy paper"*: This is Born's recollection from his AHQP interview of what Heisenberg said.

p. 115: *"formal and feeble"*: Heisenberg to Pauli, July 9, 1925, Pauli, *Briefwechsel*.

p. 115: *"looks very mystical"*: Born to Einstein, July 15, 1925, Born, Born, and Einstein, *Briefwechsel*.

p. 115: *"As Kramers has perhaps told you"*: Heisenberg to Bohr, Aug. 31, 1925, Bohr, *CW*, vol. 5, 366.

p. 115: *"An attempt is made to obtain foundations"*: Heisenberg, *Zeitschrift für Physik* 33 (1925): 879, translated in van der Waerden.

p. 116: *"has given me new* joie de vivre": Pauli to R. Kronig, Oct. 9, 1925, Pauli, *Briefwechsel*.

p. 116: *"Heisenberg has laid a large quantum egg"*: Einstein to Ehrenfest, Sept. 20, 1925, quoted by Dresden, 51.

10: The Soul of the Old System

p. 119: *The fog has begun to lift*: Moore, 187, quotes a letter from Einstein to P. Langevin, in which he said, *"Er hat eine Ecke des grossen Schleiers gelüftet,"* literally, "He [de Broglie] has lifted a corner of the great veil." But *Schleier* can also mean atmospheric haze, and the verb *lüften* is perhaps better rendered as "dispel," hence my looser translation.

p. 121: *"whitecaps" of an underlying wave field*: Moore, 187, quoting a 1926 paper by Schrödinger.

p. 121: *"a late erotic outburst in his life"*: Moore, 191; the remark is by Hermann Weyl.

p. 122: *"the concept of your paper shows real genius"* and *"I am convinced that you have made a decisive advance"*: Einstein to Schrödinger, April 16 and 26, 1926, Przibram.

p. 123: *"I know you are fond of tedious and complicated formalism"*: Born 1978, 218.

p. 124: *"awfully clever of Heisenberg"*: Born AHQP interview.

p. 125: *"the immediate task is to save Heisenberg's mechanics"*: Pauli to Kronig, Oct. 9, 1925, Pauli, *Briefwechsel*.

p. 125: *"your endless griping about Copenhagen and Göttingen"*: Heisenberg to Pauli, Oct. 12, 1925, ibid.

p. 125: *"I hardly need tell you"*: Heisenberg to Pauli, Nov. 3, 1925, ibid.

p. 126: *"was scared away if not repulsed"*: Schrödinger, *Annalen der Physik* 79 (1926): 735.

p. 126: *"extremely intricate and frighteningly abstract"*: Cassidy 1992, 213, quoting a 1927 paper by Sommerfeld.

p. 127: *"that while he understood my regrets that quantum mechanics was finished"*: Heisenberg 1971, 72.

p. 127: *"My overall impression"*: Sommerfeld to Pauli, July 26, 1926, Pauli, *Briefwechsel.*

11: I Am Inclined to Give Up Determinism

p. 130: *"the chief citadel of physics in Germany"*: Heisenberg 1989, 110.

p. 131: *he looked like a peasant boy . . . like a carpenter's apprentice*: Born 1978, 212; Pais 1991, 297.

p. 131: *the two walked through the streets to Einstein's home*: Mostly from Heisenberg 1971, ch. 5.

p. 131: *"possibly I did use that kind of reasoning"*: Ibid., 63.

p. 133: *"of the recent attempts to obtain a deeper formulation"*: Einstein to Sommerfeld, Aug. 21, 1926, in Einstein and Sommerfeld, *Briefwechsel.*

p. 133: *Einstein never felt at home among the "cool, blond Prussians"*: Frank, 113.

p. 134: *"the more I think about [it] the more repulsive I find it"*: Heisenberg to Pauli, June 8, 1926, Pauli, *Briefwechsel.*

p. 136: *Heisenberg in particular would say that the meaning of matrix elements as probabilities*: Heisenberg AHQP interview.

p. 136: *"we were so accustomed to making statistical considerations"*: Born AHQP interview.

p. 136: *"Here the whole problem of determinism arises"*: Born, *Zeitschrift für Physik* 37 (1927): 863.

p. 137: *"Quantum mechanics is very imposing"*: Einstein to Born, Dec. 4, 1926, Born, Born, and Einstein, *Briefwechsel.* "The real McCoy" is my rendition of Einstein's phrase *"der wahre Jakob,"* which is still current in some parts of Germany today. It may refer to the biblical tale in which Jacob pretends to be his brother Esau so as to gain the blessing of their father, Isaac, when he is old and blind.

p. 138: *Mrs. Bohr fussed over him with tea and cakes*: See Heisenberg's recollection in Rozental and in Heisenberg 1971, ch. 6.

p. 138: "*Are we really closer to a solution of the puzzle?*": Einstein to Sommerfeld, Nov. 28, 1926, in Einstein and Sommerfeld, *Briefwechsel.*

12: Our Words Don't Fit

p. 140: "*The conversation is almost immediately driven into philosophical questions*": Moore, 228, quoting a letter from Schrödinger to Wien, Oct. 21, 1926.

p. 140: "*pretty well spellbound*": Pais 1991, 295.

p. 141: "*since I found I couldn't express myself in French*": Dirac AHQP interview.

p. 141: "*getting the interpretation proved to be rather more difficult*": Pais 1991, 295.

p. 142: *the two men would spend hours together during the day:* See especially Heisenberg's account in Rozental.

p. 143: "*Sometimes, I had the impression that Bohr really tried to lead me onto Glatteis*": Heisenberg AHQP interview.

p. 145: "*You can look at the world with the* p-*eye*": Pauli to Heisenberg, Oct. 19, 1926, Pauli, *Briefwechsel.*

p. 149: "*our words don't fit*": Heisenberg AHQP interview.

p. 149: "*all the results in the paper are certainly correct*": Heisenberg to Pauli, May 16, 1927, Pauli, *Briefwechsel.*

p. 149: "*On the* Perceptual *Content of Quantum Theoretical Kinematics and Mechanics,*" "*On the* Physical *Content . . . ,*" *and* "*intuitive*": Cassidy 1992, 226; Pais 1991, 304; Beller, 69 and 109.

13: Awful Bohr Incantation Terminology

p. 155: "*the last word on the subject*" and "*adapting our modes of perception*": *Nature* 121 (1928), supp.: 579 (editorial comment) and 580 (Bohr). Reprinted in Bohr, *CW,* vol. 6, 52.

p. 157: "*bring us a new theory of light*": Pais 1982, 404, quoting a 1909 paper by Einstein.

p. 159: *Ehrenfest . . . inscribed on a blackboard the verse from Genesis about Babel:* Marage and Wallenborn, 154.

p. 159: *Heisenberg and Pauli professed to be unconcerned:* Pais 1991, 318, quoting a recollection by Otto Stern.

p. 160: *Bohr admitted in private that he did not entirely understand:* Ibid.; from Bohr's handwritten notes.

p. 160: *the only substantial account of the tussle between Einstein and Bohr:* Bohr's memoir was written for the Schilpp volume and is reprinted in Bohr 1961.

p. 162: *"Like a chess match . . . awful Bohr incantation terminology":* Ehrenfest to Goudsmit, Uhlenbeck, and Dieke, Nov. 3, 1927, Bohr, *CW,* vol. 6, 38 (English), 415 (German).

p. 162: *"I listened to their arguments":* Dirac in Holton and Elkana, 84.

p. 162: *"doesn't provide you with any equations":* Dirac AHQP interview.

p. 163: *"the soothing Heisenberg-Bohr philosophy":* Einstein to Schrödinger, May 31, 1928, Przibram.

14: Now the Game Was Won

p. 164: *Toward the end of the summer of 1928, a young Russian:* Gamow, 54–55.

p. 168: *"At the next meeting with Einstein":* Bohr in Schilpp, 224.

p. 169: *"followed by a court of lesser fry" and following remarks:* Rosenfeld AHQP interview.

p. 171: *"we were all quite happy":* Heisenberg AHQP interview.

p. 172: *"when I found my name in the newspapers":* Born 1968, 37.

p. 174: *"in the course of time the splendid things will separate from the hateful":* Heilbron, 154.

p. 174: *Planck was only "60 percent noble":* Fölsing, 668, quoting a letter from Einstein to F. Haber, Aug. 8, 1933.

p. 175: *After a visit to Germany in the early days of the Hitler regime:* Rosenfeld AHQP interview.

15: Life-Experience and Not Scientific Experience

p. 176: *one American visitor to Göttingen:* K. Compton, *Nature* 139 (1937): 238.

p. 178: *"I am convinced . . . that the movement to dispense with causality in physics":* This and the following remark are from Forman.

p. 180: *"Such thinking amounted to nothing more":* Gay, 79.

p. 181: *Goethe "hated mathematics":* Spengler, vol. 1, 25.

p. 181: *"The Destiny-idea demands life-experience and not scientific experience":* Ibid., 117.

p. 182: *"In the evening one goes along with what he suggests":* Einstein to Born, Jan. 27, 1920, Born, Born, and Einstein, *Briefwechsel.*

p. 183: *"I was still a classicist and not a revolutionary":* Mehra and Rechenberg, vol. 1, xxiv.

16: Possibilities of Unambiguous Interpretation

p. 187: *"In a certain sense, therefore":* Einstein's remarks are from his lecture *On the Method of Theoretical Physics* (New York: Oxford University Press, 1933). The lecture was written in German, an introductory note explains, and translated into English, not always elegantly, with the help of some Oxford physicists. Instead of the awkward "competent to comprehend the real" I have borrowed the phrase "capable of comprehending reality" from the English edition of Fölsing, 674.

p. 188: *"I cannot understand what it means to call a theory beautiful if it is not true":* Rosenfeld in Rozental, 117.

p. 189: *"Can Quantum-Mechanical Description of Physical Reality Be Considered Complete?":* A. Einstein, B. Podolsky, and N. Rosen, *Physical Review* 47 (1935): 777; reprinted in Toulmin.

p. 191: *"this onslaught came down upon us like a bolt from the blue":* Rosenfeld in Rozental, 128.

p. 191: *"lucidity and apparent incontestability":* Bohr in Schilpp, 232.

p. 191: *"a catastrophe . . . waste pen and ink"*: Pauli to Heisenberg, June 15, 1935, Pauli, *Briefwechsel*.

p. 192: *"of course, a great deal of the argument hinges"*: E. U. Condon quoted in *The New York Times*, May 4, 1935.

p. 192: *Bohr, it's not surprising to learn*: Bohr AHQP interview.

p. 192: *"What can they mean? Do you understand it?"*: Rosenfeld in Rozental, 129.

p. 192: *"The apparent contradiction in fact discloses"* and other remarks from Bohr's reply to EPR: *Physical Review* 48 (1935): 696.

p. 193: *"Rereading these passages"*: Bohr in Schilpp, 234.

p. 193: *"a final renunciation of the classical idea of causality"*: From Bohr's reply to EPR, *Physical Review* 48 (1935): 696.

p. 194: *"Bohr's principle of complementarity"*: Einstein in Schilpp, 674.

p. 196: *"appalling"* and *"high treason"*: Moore, 314, quoting a letter from Schrödinger to Einstein, March 23, 1936.

p. 196: *"It is wrong to think that the task of physics is to find out how nature is"*: Peterson.

p. 197: *"I have given up concerning myself with fundamental questions"*: Cassidy 1992, 290, quoting a letter from Heisenberg to Bohr, July 27, 1931.

p. 197: *"we cannot and should not replace these concepts by any others"*: Heisenberg 1958, 44.

p. 197: *the physicist John Bell came up with an ingeniously simple way*: The paper announcing Bell's celebrated theorem, originally published in 1964, is the second paper in Bell.

17: The No-Man's-Land Between Logic and Physics

p. 200: *"is just a way of talking about discoveries which have already been made"*: Dirac AHQP interview.

p. 200: *Bohr was at heart more of a philosopher than a physicist*: Heisenberg in Rozental, 95.

p. 201: *In 1932, Bohr spoke on "Light and Life"*: This and the following lectures are all in Bohr 1961.

p. 202: *"the concept of purpose, which is foreign to mechanical analysis"*: From the "Light and Life" lecture.

p. 203: *"whenever you come with a definite statement"*: Rosenfeld AHQP interview.

p. 207: *"in the no-man's-land that lies between logic and physics"*: Popper, 215.

p. 207: *an illuminating essay, "Causality in Contemporary Physics"*: Schlick's 1931 paper is reprinted in Toulmin.

p. 208: *an alternative interpretation of quantum mechanics*: Bohm, *Physical Review* 85 (1952): 166 and 180. For a more recent presentation, see Bohm and B. J. Hiley, *The Undivided Universe* (New York: Routledge, 1993). Beller seems to hint occasionally that she finds Bohm's version superior to the Copenhagen interpretation, while S. Goldstein, in *The Flight from Science and Reason*, ed. P. Gross, N. Levitt, and M. Lewis (New York: New York Academy of Sciences, 1996), 119, puts adherence to Copenhagen on a par with the embrace of unreason and anti-scientism. In my book *Where Does the Weirdness Go?* (New York: Basic Books, 1996), 111–21, I give some reasons why Bohm's theory is not so wonderful either.

p. 208: *"That way seems too cheap to me"*: Einstein to Born, May 12, 1952, Born, Born, and Einstein, *Briefwechsel*.

18: Anarchy at Last

p. 210: *"the more precisely the media measures individual events in a war"*: Tony Blankley, *Washington Times*, April 3, 2003.

p. 211: *"formulas, diagrams"*: Gore Vidal's essay, *New York Review of Books*, July 17, 1976, and see letters in the Oct. 28 issue.

p. 212: The West Wing: Season 5, episode 18, "Access."

p. 215: *"He found himself in a land where no one had ever penetrated before"*: Adams, 457–58.

p. 216: *"the typical physicist feels"*: See Bell, 28n8; paper written with M. Nauenberg.

Postscript

p. 220: *"I don't like your kind of physics"*: Heisenberg AHQP interview.

p. 221: *"in between he was a pleasant man"*: Margrethe Bohr AHQP interview.

BIBLIOGRAPHY

Of the vast literature on quantum theory and its history I have read only a small fraction, and I include in this list only that still smaller fraction that I have found particularly illuminating.

Adams, H. *The Education of Henry Adams*. Boston: Houghton Mifflin, 1961.

Bell, J. S. *Speakable and Unspeakable in Quantum Mechanics*. Cambridge, U.K.: Cambridge University Press, 1987.

Beller, M. *Quantum Dialogue: The Making of a Revolution*. Chicago: University of Chicago Press, 1999.

Bohr, N. *Atomic Physics and Human Knowledge*. New York: Science Editions, 1961. (Includes "Discussion with Einstein on Epistemological Problems in Atomic Physics," from Schilpp 1949.)

———. *Collected Works*. Ed. L. Rosenfeld. 11 vols. Amsterdam: North-Holland, 1972–87.

Born, M. *My Life and My Views*. New York: Charles Scribner's Sons, 1968.

———. *My Life: Recollections of a Nobel Laureate*. New York: Charles Scribner's Sons, 1978.

Born, M., H. Born, and A. Einstein. *Briefwechsel, 1916–1955. Kommentiert von Max Born*. Munich: Nymphenburger, 1969. In English: *The Correspondence Between Albert Einstein and Max and Hedwig Born, 1916–1955, with Commentaries by Max Born*. Trans, I. Born. New York: Walker, 1971.

Cassidy, D. C. "Answer to the Question: When Did the Indeterminacy Principle Become the Uncertainty Principle?" *American Journal of Physics* 66 (1998): 278.

———. *Uncertainty: The Life and Science of Werner Heisenberg*. New York: W. H. Freeman, 1992.

Dresden, M. *H. A. Kramers: Between Tradition and Revolution*. New York: Springer-Verlag, 1987.

Einstein, A., and A. Sommerfeld. *Briefwechsel*. Ed. A. Hermann. Basel, Switzerland: Schwabe, 1968.

Enz, C. P. *No Time to Be Brief: A Scientific Biography of Wolfgang Pauli*. New York: Oxford University Press, 2002.

Eve, A. S. *Rutherford*. Cambridge, U.K.: Cambridge University Press, 1939.

Fölsing, A. *Albert Einstein*. New York: Viking, 1997.

Forman, P. "Weimar Culture, Causality, and Quantum Theory, 1918–1927: Adaptation by German Physicists and Mathematicians to a Hostile Intellectual Environment." *Historical Studies in the Physical Sciences* 3 (1971): 1.

Frank, P. *Einstein: His Life and Times*. New York: A. A. Knopf, 1953.

Gamow, G. *Thirty Years That Shook Physics: The Story of Quantum Theory*. New York: Dover, 1985.

Gay, P. *Weimar Culture: The Outsider as Insider*. New York: Harper & Row, 1968.

Gillispie, C. C., ed. *Dictionary of Scientific Biography*. New York: Scribner, 1970–89.

Greenspan, N. T. *The End of the Certain World: The Life and Science of Max Born*. New York: Basic Books, 2005.

Heilbron, J. L. *The Dilemmas of an Upright Man: Max Planck as Spokesman for German Science*. Berkeley: University of California Press, 1986.

Heisenberg, W. *Encounters with Einstein*. Princeton, N.J.: Princeton University Press, 1989.

———. *Physics and Beyond: Encounters and Conversations*. New York: Harper & Row, 1971.

———. *Physics and Philosophy*. New York: Harper, 1958.

Hendry, J. "Weimar Culture and Quantum Causality." *History of Science* 18 (1980): 155.

Holton, G., and Y. Elkana, eds. *Albert Einstein: Historical and Cultural Perspectives*. New York: Dover, 1997.

Kilmister, C. W., ed. *Schrödinger: Centenary Celebration of a Polymath*. New York: Cambridge University Press, 1987.

Kragh, H. "The Origin of Radioactivity: From Solvable Problem to Unsolved Non-problem." *Archive for the History of the Exact Sciences* 50 (1997): 331.

———. *Quantum Generations: A History of Physics in the Twentieth Century*. Princeton, N.J.: Princeton University Press, 1999.

Kuhn, T. S. *Black-Body Theory and the Quantum Discontinuity, 1894–1912*. Chicago: University of Chicago Press, 1978.

Laqueur, W. *Weimar: A Cultural History*. New York: G. P. Putnam's Sons, 1974.

Lindley, D. *Boltzmann's Atom: The Great Debate That Launched a Revolution in Physics*. New York: Free Press, 2001.

Marage, P., and G. Wallenborn. *The Solvay Councils and the Birth of Modern Physics*. Boston: Birkhäuser, 1999.

Mehra, J., and H. Rechenberg. *The Historical Development of Quantum Theory*. 6 vols. New York: Springer, 1982–2001.

Meyenn, K. von, and E. Schucking. "Wolfgang Pauli." *Physics Today*, Feb 2001.

Mommsen, H. *The Rise and Fall of Weimar Democracy*. Trans. E. Forster and L. E. Jones. Chapel Hill: University of North Carolina Press, 1996.

Moore, W. *Schrödinger: Life and Thought*. New York: Cambridge University Press, 1989.

Nelson, E. *Dynamical Theories of Brownian Motion*. Princeton, N.J.: Princeton University Press, 1967. (Second edition, 2001, available at www.math.princeton.edu/~nelson/books.html.)

Nye, M. J. *Molecular Reality: A Perspective on the Scientific Work of Jean Perrin*. New York: History of Science Library, 1972.

———, ed. *The Question of the Atom: From the Karlsruhe Congress to the First Solvay Conference, 1860–1911*. Los Angeles: Tomash, 1984.

Pais, A. *Inward Bound: Of Matter and Forces in the Physical World*. New York: Oxford University Press, 1986.

———. *Niels Bohr's Times in Physics, Philosophy, and Polity*. New York: Oxford University Press, 1991.

———. *Subtle Is the Lord . . . : The Science and the Life of Albert Einstein*. New York: Oxford University Press, 1982.

Pauli, W. *Wissenschaftlicher Briefwechsel mit Bohr, Einstein, Heisenberg u. A.* Ed. A. Hermann and K. von Meyenn. Vol. 1, *1919–1929*. New York: Springer, 1979.

Peterson, A. "The Philosophy of Niels Bohr." *Bulletin of the Atomic Scientists*, Sept. 1963, 8.

Petruccioli, S. *Atoms, Metaphors, and Paradoxes: Niels Bohr and the Construction of a New Physics*. New York: Cambridge University Press, 1993.

Popper, K. *The Logic of Scientific Discovery*. New York: Basic Books, 1958.

Przibram, K., ed. *Brief zur Wellenmechanik: Schrödinger, Planck, Einstein, Lorentz*. Vienna: Springer, 1963. In English: *Letters on Wave Mechanics*. Trans. M. J. Klein. New York: Philosophical Library, 1967.

Quinn, S. *Marie Curie*. Reading, Mass.: Addison-Wesley, 1995.

Rozental, S., ed. *Niels Bohr: His Life and Work as Seen by His Friends and Colleagues*. Amsterdam: North-Holland, 1968.

Schilpp, P. A., ed. *Albert Einstein: Philosopher-Scientist*. Evanston, Ill.: Library of Living Philosophers, 1949.

Segrè, E. *From X-Rays to Quarks: Modern Physicists and Their Discoveries*. San Francisco: W. H. Freeman, 1980.

Spengler, O. *The Decline of the West*. Trans. C. F. Atkinson. 2 vols. New York: A. A. Knopf, 1926–28.

Stachura, P. D. *Nazi Youth in the Weimar Republic*. Santa Barbara, Calif.: Clio, 1975.

Stuewer, R. K. *The Compton Effect: Turning Point in Physics*. New York: Science History Publications, 1975.

Toulmin, S., ed. *Physical Reality: Philosophical Essays on Twentieth-Century Physics*. New York: Harper & Row, 1970.

Waerden, B. van der, ed. *Sources of Quantum Mechanics*. New York: Dover, 1967.

INDEX

Printed in the United States
by Baker & Taylor Publisher Services